Decomposition-based Evolutionary Optimization in Complex Environments

Decomposition-based Evolutionary Optimization in Complex Environments

Juan Li
Bin Xin
Jie Chen

Beijing Institute of Technology, China

World Scientific

NEW JERSEY · LONDON · SINGAPORE · BEIJING · SHANGHAI · HONG KONG · TAIPEI · CHENNAI · TOKYO

Published by

World Scientific Publishing Co. Pte. Ltd.

5 Toh Tuck Link, Singapore 596224

USA office: 27 Warren Street, Suite 401-402, Hackensack, NJ 07601

UK office: 57 Shelton Street, Covent Garden, London WC2H 9HE

British Library Cataloguing-in-Publication Data

A catalogue record for this book is available from the British Library.

DECOMPOSITION-BASED EVOLUTIONARY OPTIMIZATION IN
COMPLEX ENVIRONMENTS

ISBN 978-981-121-898-9 (hardcover)
ISBN 978-981-121-899-6 (ebook for institutions)
ISBN 978-981-121-900-9 (ebook for individuals)

For any available supplementary material, please visit
https://www.worldscientific.com/worldscibooks/10.1142/11788#t=suppl

Desk Editor: Amanda Yun

Typeset by Stallion Press
Email: enquiries@stallionpress.com

Preface

Life inevitably involves making decisions and searching for optima in complex environments. The majority of real-world applications contain two or more objectives to be optimized. It is natural to want all of these objectives to be optimal. However, multiple objectives are usually incomparable, which means that they have different scales in most cases. What's more, these objectives tend to contradict each other (at least partially, if not fully) and we cannot improve a certain objective without deteriorating the performance of the remaining objectives. Thus, the only way to optimize multiple conflicting objectives is to make a trade-off among them. The optimization problems that are characterized by the simultaneous optimization of multiple conflicting objectives are known as multi-objective optimization problems (MOPs). The area of multiple objective optimization has developed rapidly. The MOPs to be solved vary from weapon-target assignment in the field of military operations research, to redundancy allocation in the design of a complex system. The essence of optimizing MOPs is to achieve a trade-off among multiple conflicting objectives. MOPs usually contain an infinite number of feasible solutions. Hwang and colleagues classified approaches for multi-objective optimization as the following three categories according to the participation of the decision maker in the solution process: *a priori* methods, interactive methods and *a posteriori* methods. The evolutionary algorithm (EA) is a type of classical stochastic optimization method and has been proved to be capable of dealing with complicated problems. Multi-objective evolutionary algorithms (MOEAs) focus on the posterior cases, which means designing efficient methods to obtain a set of non-dominated solutions that can be presented to decision makers to select from based on their preferences. In the literature, a category of MOPs that involves more than three objectives is termed as the many-objective optimization problem (MaOP). MaOPs appear widely in the field of industrial and engineering design, such as water resource engineering problems, industrial scheduling problems, molecular design problems, and automotive engine calibration problems. The high-dimensional objectives pose serious difficulties to MOEAs originally designed for MOPs.

Uncertainties widely exist in real-world applications, including operations research, management science, information science, system science, and engineering applications. Generally, existing techniques on dealing with uncertain optimization problems (UOPs) are either based on the knowledge of decision makers' (DMs) preferences on uncertainties or are only suitable for limited types of DMs' preferences. These methods can deal with only one type of preference. Actually, it is difficult for a DM to provide his/her preferences accurately, since the DM usually knows little about uncertainties before making decisions. Besides, it may be obtrusive or even risky to incorporate the preferences of a DM to handle uncertainties when he/she does not have sufficient knowledge about the problem. Besides, in many practical applications, problems are not only characterized by multiple objectives to be optimized but also influenced by uncertainties as a consequence of uncontrollable factors. These problems are referred to as uncertain multi-objective problems (UMOPs). Standard MOEAs are inadequate to deal with UMOPs. The key point of optimizing UMOPs efficiently is to eliminate detrimental effects induced by noise and strike a balance between convergence and diversity from the aspects of objective space and decision space. Many phenomena are of the multi-objective and uncertain nature, which is why we need tools for handling several conflicting and uncertain objectives in complex environments.

Decomposition is an efficient and prevailing strategy in the field of traditional mathematical programming. Several methods for constructing aggregation functions can be found in the literature. The most popular ones among them include the weighted sum approach and Tchebycheff approach. Recently, the boundary intersection method and the ε-constraint method have also attracted a lot of attention. The success of decomposition has been witnessed by the multi-objective evolutionary algorithm MOEA/D and its variants. In decomposition-based methods, an MOP is decomposed into a number of scalar subproblems by using various scalarizing functions. Each individual solution in the population is associated with a subproblem. A neighborhood relationship among all the subproblems is defined based on the distances of their associated vectors. Then, each subproblem is optimized by only using information from its several neighboring subproblems, since two neighboring subproblems should have similar optimal solutions.

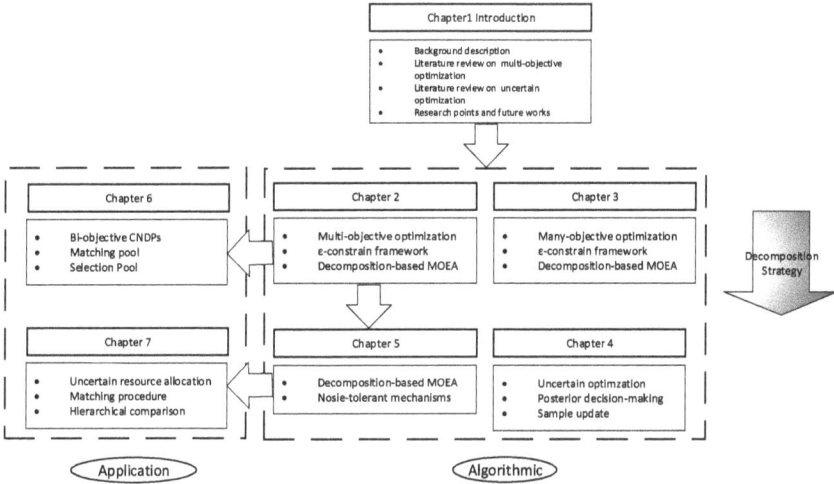

Fig. 0.1   Research structure of this book.

In this book, we concentrate on decomposition-based evolutionary optimization in complex environments. The content of this book can be divided into seven chapters. The main structure of this book is presented in Fig. 0.1. To be specific, Chapter 1 introduces the basic definition and connotation of multi-objective optimization and uncertain optimization, and presents a brief overview of the research situation of multi-objective algorithms and uncertainty handling mechanisms. Chapter 2 puts forward the authors' research on applying the decomposition strategy and incorporates the ε-constraint method into the decomposition strategy, which gives birth to a new decomposition-based multi-objective evolutionary algorithm (DMOEA) with the ε-constraint framework (DMOEA-εC). Furthermore, in Chapter 3, the authors further investigate the performance of DMOEA-εC on dealing with MaOPs. In order to overcome ineffectiveness induced by the exponential number of upper bound vectors and to strike a balance between convergence and diversity, four improved strategies are put forward. By embedding these strategies into a DMOEA-εC, an MOEA named DMaOEA-εC, which are applicable for both MOPs and MaOPs are developed. Chapter 4 is devoted to the applications of the decomposition-based strategy in the field of uncertain optimization. It treats UOPs in an *a posteriori* manner and proposes an *a posteriori* decision-making framework that covers several common uncertain models. Furthermore, a subproblems co-solving evolutionary algorithm for UOPs, i.e., S-CoEA, is proposed by incorporating the weighting method into a decomposition strategy. Chapter 5 describes the work of utilizing DMOEAs to solve UMOPs. To be specific, experiments are firstly conducted to examine the impact of noisy environments on MOEAs. Then, four noise-handling techniques based upon the analyses of empirical results are proposed to enhance the performance of standard DMOEAs on dealing with UMOPs. Chapter 6 describes the application of DMOEAs on solving the bi-objective critical node detection problem (Bi-CNDP). We first prove the NP-hardness of this problem for general graphs and the existence of a polynomial algorithm for constructing the ε-approximated Pareto front for Bi-CNDPs on trees. Besides, different approaches of determining the mating pool and the replacement pool are proposed for DMOEAs. Chapter 7 demonstrates the application of DMOEAs on uncertain bi-objective resource allocation problems. It transforms uncertain optimization problems into deterministic ones by using the conditional value-at-risk (CVaR) measure and proposes different forms of matching procedures and hierarchical comparison strategies to further improve the performance of DMOEAs in a simple and efficient way.

Juan Li, Bin Xin and Jie Chen

Acknowledgments

This book would not exist without the help and care of many people. The authors wish to thank all the researchers of the school of Automation and the school of Mechatronical Engineering, Beijing Institute of Technology, the State Key Laboratory of Intelligent Control and Decision of Complex Systems and the Beijing Advanced Innovation Center for Intelligent Robots and Systems for reading and commenting on draft versions of the book. Juan Li has benefited from a grant from the China Scholarship Council and the China Postdoctoral Science Foundation. Bin Xin and Jie Chen acknowledge the support of the National Natural Science Foundation of China under Grants 61822304 and 61673058, the NSFC-Zhejiang Joint Fund for the Integration of Industrialization and Informatization under Grant U1609214, the Foundation for Innovative Research Groups of the National Natural Science Foundation of China under Grant 61621063, the Projects of Major International (Regional) Joint Research Program NSFC under Grant 61720106011, and the Peng Cheng Laboratory.

In addition, Juan Li would like to offer special thanks to Professors Panos M. Pardalos and Stan Uryasev of the University of Florida for their careful guidance in scientific research. Communicating with them has broadened her academic thinking and horizons. She also thanks her parents and family members, who have been standing silently behind her, for their spiritual and financial support which allowed her to complete her studies. Last but not least, Juan Li would also like to thank those she met and shared experiences with over the years, for the stories that have made her grow and mature, particularly the motto "Try to be better", which has driven her research, saved her from depression, warmed her during periods of loneliness, and encouraged her in the midst of struggles. She hopes that their pursuit of excellence will not be consumed by worldly trivia, and their dreams not fade with time.

About the Authors

Juan Li received her B.S. degree in Statistics and Ph.D. degree in Control Science and Engineering, both from the Beijing Institute of Technology (BIT), Beijing, China, in 2013 and 2019, respectively. She is currently a post-doctorate fellow with the School of Mechatronical Engineering at BIT. Her current research interests include multi-objective evolutionary optimization, combinatorial optimization and uncertain optimization, and swarm intelligence.

Bin Xin received his B.S. degree in Information Engineering and Ph.D. degree in Control Science and Engineering, both from BIT, in 2004 and 2012, respectively. He is currently a professor with the School of Automation at BIT. His current research interests include search and optimization, evolutionary computation, unmanned systems, and multi-agent systems.

Jie Chen received his B.S., M.S., and Ph.D. degrees in Control Theory and Control Engineering from BIT in 1986, 1996, and 2001, respectively. From 1989 to 1990, he was a visiting scholar in the California State University, US. From 1996 to 1997, he was a research fellow in the School of Engineering, University of Birmingham, UK. He is currently a professor of Control Science and Engineering at BIT and the Tongji University, China. He is also a member of the Chinese Academy of Engineering, an IEEE Fellow and an IFAC Fellow. His main research interests include multi-objective optimization and decision-making in complex systems, intelligent control, and nonlinear control. He has co-authored 4 books and more than 200 research papers.

Contents

Chapter 1

Introduction

1.1 Uncertain and Multi-objective Optimization

In real-world applications, there are usually many decision-making problems that contain two or more objectives to be optimized. For example, the goal of the portfolio investment problem is to maximize the profit of investment and minimize its risk at the same time. The weapon-target assignment problem aims to maximize the expected damage to targets and minimize the cost of ammunition cost. The aim of the redundancy allocation problem is to maximize the overall reliability of the system and minimize the cost of redundancy components simultaneously. In most cases, these multiple objectives are incomparable, which means that these objectives have different scales and thus are incomparable. What's more, these objectives are contradicted with each other and we cannot improve a certain objective without deteriorating the performance of remaining objectives. Thus, the only way to optimize multiple conflicting objectives is to make a trade-off among them. The optimization problems that are characterized by multiple conflicting objectives to be optimized simultaneously are known as multi-objective optimization problems (MOPs) [Miettinen (1999)]. Solving MOPs has become an interesting area of research during the past few decades in both the theoretical and engineering fields. MOPs widely exist in the field of decision science, computer science, design of complex systems, planning, and so on.

The essence of optimizing MOPs is to achieve a trade-off among multiple conflicting objectives. Traditional approaches convert an MOP into a single objective optimization problem (SOP), and then adopt existing single objective optimizers to handle it. The key of these methods lies in how to transfer an MOP into a SOP. Commonly used aggregation functions include the weighted sum function, Tchebycheff function, ε-constraint method, objective programming and max-min method [Miettinen (1999)]. The traditional method can inherit well-researched mechanisms of single objective optimization, which shows superiority and attraction. However, the traditional method has the following drawbacks. Firstly, some aggregation functions such as the weighted sum function are sensitive to the shape of the Pareto front (PF) and cannot deal with the concave part of the PF. Secondly, it is difficult to obtain the knowledge (such as weight vectors, upper bound vectors, and so on) needed for the transformation. What's more, traditional methods should be conducted multiple times in order to obtain a set of Pareto optimal solutions, namely the Pareto optimal set. The results obtained may be inconsistent with others due to the independence of multiple runs, which makes it difficult to make decisions efficiently. Finally, multiple runs decrease the computational efficiency.

The evolutionary algorithm (EA) is a type of classical stochastic optimization method and has been proved to be capable of dealing with complicated SOPs. Since the mid-1980s, EAs have been adopted in the field of multi-objective optimization, replaced traditional approaches, and opened a new way for multi-objective optimization. EAs do not require gradient information, convex property, differentiability, and continuity of objective functions, which extends the range of application. Besides, EAs are insensitive to the shape and continuity of PFs, which make it able to approximate PFs with various shapes. Multi-objective evolutionary algorithms (MOEAs) [Zhou *et al.* (2011)] have become useful tools for MOPs and a hot research topic in the field of evolutionary computation.

MOPs usually contain multiple, even infinite number of feasible solutions, thus decision makers should be able to select the most preferred one via a certain search process. Hwang and Masud (1979) classified approaches for multi-objective optimization as the following three categories, according to the participation of the decision maker in the solution process:

(1) Methods where *a priori* articulation of preference information is used;
(2) Methods where interactive articulation of preference information is used;
(3) Methods where *a posteriori* articulation of preference information is used;

The first group is called *a priori* methods in which a decision maker (DM) must specify her or his preferences, hopes and opinions before the solution process. The second group is called the interactive method. The interest devoted to this class can be explained by the fact that, assuming the DM has enough time and capabilities for co-operation, interactive methods can be presumed to produce the most satisfactory results. The last group is called *a posteriori* methods, which can also be called methods for generating Pareto optimal solutions. After the Pareto optimal set (or a part of it) has been generated, it is presented to the DM to select the most preferred solution among the alternatives. The research on MOEAs mostly focus on the posterior cases, which means designing efficient methods to obtain a set of non-dominated solutions that can be presented to DMs to select from according to their preferences.

In the literature, a category of MOPs that involve more than three objectives is termed as the many-objective optimization problem (MaOP) [Roy *et al.* (2015)]. MaOPs appear widely in the field of industrial and engineering design, such as water resource engineering problems [Kasprzyk *et al.* (2009)], industrial scheduling problems [Slflow *et al.* (2007)], molecular design problems [Kruisselbrink *et al.* (2009)], and automotive engine calibration problems [Lygoe *et al.* (2013)]. Many-objective optimization has been gaining attention in the evolutionary multi-objective optimization community during recent years.

The high-dimensional objectives pose serious difficulties to MOEAs which are originally designed for MOPs. First and foremost, the proportion of non-dominated solutions in a population increases rapidly with the number of objectives [Ishibuchi *et al.* (2008)]. This decreases the effectiveness of the Pareto-based selection scheme, since Pareto dominance relationships fail to distinguish individuals. Then, the diversity-based selection mechanisms play a leading role in both mating and environmental selection, which slows down the evolutionary process and results in solutions that may be distributed uniformly but away from the desired PF in the objective space [Hadka and Reed (2012)]. Due to this challenge, most classical Pareto-based MOEAs [Zitzler *et al.* (2001)] noticeably deteriorate their performance when solving MaOPs [Ishibuchi *et al.* (2016)]. In addition to the difficulty in convergence, maintaining diversity is another challenge in many-objective optimization. The conflict between convergence and diversity becomes aggravated in a high-dimensional space and the similarity between solutions is more difficult to estimate [Purshouse and Fleming (2007)]. Most of the current diversity management operations cannot strengthen the selection pressure towards the PF and may even impede the evolutionary process to a certain extent. Next, the number of points needed to represent the desired PF becomes very large [Ishibuchi *et al.* (2016)], but the population size used in MOEAs cannot be arbitrarily large due to limited computational resources. In the high-dimensional objective space, limited number of solutions are likely to be far away from each other.

This might lead to the inefficiency of generating offsprings, since reproduction operators usually produce offsprings that are far away from their parents in a high-dimensional space [Ishibuchi *et al.* (2014)]. Besides, it is well-known that the computational complexity for calculating some performance metrics, such as hypervolume (HV) [Zitzler and Thiele (1999)], grows exponentially with the increasing number of objectives [While *et al.* (2006)]. This poses difficulties on applying indicator-based MOEAs to MaOPs. Finally, it is highly challenging to visualize a high-dimensional PF. Therefore, the selection of a preferred solution becomes a difficult task for decision makers. The above-mentioned difficulties make it difficult to apply algorithms, which are originally designed for MOPs, to MaOPs. Therefore, it is necessary to design new approaches for MaOPs according to their characteristics or improve the original MOEAs to make them applicable for MaOPs.

Uncertainties widely exist in real-world applications, for example, operations research, management science, information science, system science, and engineering applications. Generally, existing techniques on dealing with uncertain optimization problems (UOPs) are either based on the knowledge of DMs' preferences on uncertainties or only suitable for limited types of DMs' preferences. These methods can deal with only one type of preference at a time. Actually, it is difficult for a DM to provide his/her preferences accurately, since the DM usually knows little about uncertainties before making decisions. What's more, it may be obtrusive or even risky to incorporate the preferences of a DM to handle uncertainties when he/she does not have sufficient knowledge about the problem. Besides, in many practical applications, problems are not only characterized by multiple objectives to be optimized, but are also influenced by uncertainties as a consequence of uncontrollable factors. These problems to be considered are referred to as uncertain multi-objective problems (UMOPs). Standard MOEAs are inadequate to deal with UMOPs. The key point of optimizing UMOPs efficiently is to eliminate detrimental effects induced by noise and strike a balance between convergence and diversity from the aspects of objective space and decision space.

1.2 Mathematical Formulations of the Uncertain and Multi-objective Optimization Problem

This section presents basic definitions evolved in this book. Subsections 1.2.1 and 1.2.2 describe concepts related to MOPs and UOPs, respectively.

1.2.1 *Multi-objective Optimization Problems*

Many real-world practical problems have two or more objectives that are usually conflicted. These problems to be handled are named multi-objective optimization problems (MOPs). An MOP can be stated as follows:

$$\begin{aligned} \text{minimize} \quad & \mathbf{F}(\mathbf{x}) = (f_1(\mathbf{x}), f_2(\mathbf{x}), \dots, f_m(\mathbf{x})) \\ \text{subject to} \quad & \mathbf{x} = (x_1, x_2, \dots, x_n) \in \Omega, \end{aligned} \quad (1.1)$$

where $\mathbf{x} = (x_1, x_2, \dots, x_n)$ is the decision vector. It belongs to the nonempty feasible set Ω. $\mathbf{F} : \Omega \to R^m$ consists of $m (\geq 2)$ objective functions $f_i : R^n \to R$, $i = 1, \dots, m$. When $m \geq 4$, this problem is termed as many-objective optimization problems (MaOPs). The objective functions in (1.1) contradict each other, and no single solution optimizes them simultaneously. Otherwise, it will be deteriorated into a single objective problem (SOP). Any improvement in one objective of a Pareto optimal point must lead to deteriorations in at least one other objective. The Pareto optimality is a trade-off among these conflicted objectives [Deb *et al.* (2006); Miettinen (1999)], and the exact definition of Pareto dominance and Pareto optimality is firstly put forward by Edgeworth and Pareto.

Definition 1.1 (Pareto Dominance). A vector $\mathbf{u} = (u_1, u_2, \dots, u_m)$ dominates a vector $\mathbf{v} = (v_1, v_2, \dots, v_m)$, denoted as $\mathbf{u} \prec \mathbf{v}$, if and only if $u_i \leq v_i$ for every $i \in \{1, \dots, m\}$ and $u_j < v_j$ strictly holds for at least one index $j \in \{1, \dots, m\}$.

Definition 1.2 (Pareto Optimality). A decision vector $x^* \in \Omega$ is Pareto optimal if there does not exist another decision vector $y \in \Omega$ such that $\mathbf{F}(y) \prec \mathbf{F}(x^*)$. The set of all the Pareto optimal points is called the Pareto set (PS), i.e., $PS = \{x \in \Omega | \neg \exists y \in \Omega, \mathbf{F}(y) \prec \mathbf{F}(x^*)\}$. The set of all the Pareto optimal objective vectors is called the Pareto front (PF), i.e., $PF = \{\mathbf{F}(x) | x \in PS\}$.

Definition 1.3 (Ideal Point $\mathbf{z}^*$). The component z_i^* of the ideal point $\mathbf{z}^* \in R^m$ is obtained by minimizing each of the objective functions individually subject to the constraints, that is, by solving the following problems:

$$\begin{aligned} \text{minimize} \quad & f_i(\mathbf{x}) \\ \text{subject to} \quad & \mathbf{x} \in \Omega \end{aligned} \quad \text{for } i = 1, \dots, m. \quad (1.2)$$

Definition 1.4 (Nadir Point $\mathbf{z}^{\mathbf{nad}}$). The nadir point is the upper bound of the PF. Each element z_i^{nad} of the nadir point $\mathbf{z}^{\mathbf{nad}} \in R^m$ is defined as $z_i^{nad} = \max\{f_i | \mathbf{F} = (f_1, f_2 \ldots, f_m) \in PF\}$ for $i = 1, \ldots, m$.

Definition 1.5 (Proper Pareto Optimality). A decision vector $x^* \in \Omega$ is proper Pareto optimal if it is Pareto optimal and if there is some real number $M > 0$ such that for each f_i and each $\mathbf{x} \in \Omega$ satisfying $f_i(\mathbf{x}) < f_i(\mathbf{x}^*)$, there exists at least one f_j such that $f_j(\mathbf{x}^*) < f_j(\mathbf{x})$ and $\frac{f_i(\mathbf{x}^*) - f_i(\mathbf{x})}{f_j(\mathbf{x}) - f_j(\mathbf{x}^*)} \leq M$. An objective vector is properly Pareto optimal if the decision vector corresponding to it is properly Pareto optimal.[1]

Figure 1.1(a) gives an illustration of the concept of Pareto dominance relationship, in which solution A dominates B, B dominates solutions C and D, and solutions E and B are non-dominated. Besides, we can see the region that is dominated by solution B, the region that dominates B, and the region that is non-dominated with B. To be specific, solutions located in the top right corner of the figure are with inferior performance compared with B. The lower left corner is the region that has better performance than B. Solutions located in the remaining two sub-regions are non-dominated with B. According to Definition 1.2, Fig. 1.1(b) presents the non-dominated

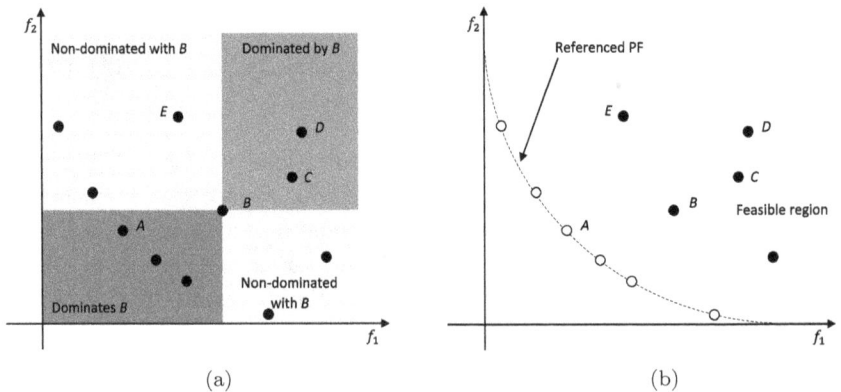

(a)   (b)

Fig. 1.1   (a) An illustration of the concept of Pareto dominance and (b) Pareto optimal solutions in the objective space.

---

[1]The proper Pareto optimality is more strict than the Pareto optimality. A solution is properly Pareto optimal if there is at least one pair of objectives for which a finite decrement in one objective is possible only at the expense of some reasonable increment in the other objective.

solutions (denoted as hollow circles) in a population, in which the dotted line means the PF. In addition to the commonly used Pareto dominance relationship, other types of Pareto dominance can be found in [Miettinen (1999)], such as strong Pareto dominance, weak Pareto dominance, local Pareto dominance, global Pareto dominance, and so on.

With an aim of comparing the performance of various MOEAs, researchers put forward a series of benchmarks including continuous, discrete and real-world problems for multi-objective optimization. The ZDT test instances [Zitzler *et al.* (2000)], tri-objective DTLZ test instances [Deb *et al.* (2002b)], UF test suites [Zhang *et al.* (2009b)] (which are part of the CEC2009 MOP test instances), LZ test suites [Li and Zhang (2009)] (with complicated PS shapes), and bi-objective WFG test suites [Huband *et al.* (2005)] (with complicated PF shapes) are commonly used continuous benchmarks. The discrete test instances cover the multi-objective 0/1 knapsack problem (MOKP) [Shim *et al.* (2012)], multi-objective traveling salesman problem (MTSP) [Lust and Teghem (2010)], and so on. The real-world applications include the weapon-target assignment problem, the critical node detection problem and the resource allocation problem.

It is very important to evaluate the performance of non-dominated solutions obtained by different MOEAs. In previous research, researchers did not realize the importance of performance metrics and thus did not propose systematic metrics. Recently, researchers noticed this problem and conducted research in this field. The true PF of most benchmark problems can be obtained and denoted as P^*. However, as to some real-world problems, since it is difficult to obtain the P^*, we can try to obtain a referenced PF by using problem-specific knowledge. Otherwise, we can compare the relative performance of two algorithms without using the true PF. It is usually desired that an MOEA achieve good performance in terms of the following three aspects:

(1) The set of obtained non-dominated solutions approximated to be as near as possible;
(2) The distribution of obtained non-dominated solutions to be as uniform as possible;
(3) The coverage of obtained non-dominated solutions to be as wide as possible, which means obtained non-dominated solutions should cover the range of each objective.

Among them, (1) indicates the convergence; (2) and (3) both represent the diversity. According to the three requirements, the performance metrics can be grouped into three categories: measuring only convergence, measuring only diversity, and measuring both convergence and diversity. Besides, metrics can be classified into unitary and binary ones based on the dependence of P^*. The unitary metrics means compare the obtained optimal set with the true PF P^*, while binary measures compare the relative performance of two MOEAs. More information on performance metrics can be found in [Zitzler *et al.* (2003)]. Next, we will give a brief introduction of four performance metrics, which will be used in the following part of this book.

(1) Inverted Generational Distance [Zhou *et al.* (2005)] (*IGD*). The *IGD* metric measures the average distance from a set of uniformly distributed Pareto optimal points over the PF P^* to the approximation set P. It can be formulated as:

$$IGD(P^*, P) = \frac{\sum_{x^* \in P^*} d(x^*, P)}{|P^*|}, \tag{1.3}$$

where $d(x^*, P)$ is the minimal Euclidean distance between x^* and any point in P, and $|P^*|$ is the cardinality of P^*. If $|P^*|$ is large enough to represent the PF very well, $IGD(P^*, P)$ could measure both diversity and convergence of P in a sense. A smaller IGD value indicates a better P.

(2) Hypervolume [Zitzler and Thiele (1999)] (*HV*). The *HV* metric measures the size of the objective space dominated by the solutions in P and bounded by the reference point $\mathbf{r}$. It is defined as:

$$HV(P, \mathbf{r}) = VOL\left(\bigcup_{x \in P} [f_1(x), r_1] \times \cdots \times [f_m(x), r_m]\right), \tag{1.4}$$

where $\mathbf{r} = (r_1, \ldots, r_m)$ is a reference point in the objective space dominated by any Pareto optimal point, and $VOL(\cdot)$ is the Lebesgue measure. A larger HV value implies a better P.

(3) Additive ϵ-indicator [Zitzler *et al.* (2003)] ($I_{\epsilon+}$). The $I_{\epsilon+}$ gives the factor by which an approximation set P is worse than the PF P^* with respect to all objectives. It is formulated as:

$$I_{\epsilon+}(P, P^*) = \inf_{\epsilon \in R^+} \{\forall z^2 \in P^*, \exists z^1 \in P : z^1 \succeq_{\epsilon+} z^2\}, \tag{1.5}$$

where $z^1 \succeq_{\epsilon+} z^2$ if and only if $\forall i \in \{1, 2, \ldots, m\} : z_i^1 \leq \epsilon + z_i^2$. It measures the convergence of the approximation set P. A smaller $I_{\epsilon+}$ value indicates a better P.

(4) Averaged Hausdorff distance [Esquivel *et al.* (2012)] (Δ_p). The Δ_p metric measures the averaged Hausdorff distance between a set of uniformly distributed Pareto optimal points over the PF P^* and an approximation set P. It is composed of the generalized generational distance (GD_p) [Jin *et al.* (2001)] and the generalized inverted generational distance (IGD_p) [Liu *et al.* (2014)], which can be formulated as:

$$\Delta_p(P^*, P) = \max\{IGD_p(P^*, P), GD_p(P^*, P)\}, \qquad (1.6)$$

where $GD_p(P^*, P) = (\frac{1}{|P^*|} \sum_{x^* \in P^*} d(x^*, P)^p)^{1/p}$ and $IGD_p(P^*, P) = (\frac{1}{|P|} \sum_{x^* \in P} d(x^*, P^*)^p)^{1/p}$ are generalized GD and IGD metrics, respectively. $d(a, B)$ means the minimum Euclidean distance between the point a and any point in B. $|\cdot|$ denotes the cardinality of any set. Δ_p can detect outliers in candidate solutions. A smaller Δ_p indicates a better P.

1.2.2 *Uncertain Optimization Problems*

Many real-world optimization problems, such as system designing and planning and scheduling problems are often characterized by the disruptive presence of uncertainties induced by incomplete, unobtainable and unquantifiable information. These problems to be handled are named uncertain optimization problems (UOPs). A UOP can be stated as follows[2]:

$$
\begin{aligned}
\text{minimize} \quad & f(\mathbf{x}, \xi) \\
\text{subject to} \quad & g_j(\mathbf{x}, \xi) \leq 0, j = 1, \ldots, p,
\end{aligned}
\qquad (1.7)
$$

where $\mathbf{x}$ is the decision vector, ξ denotes the uncertain parameter, and different values of ξ represent different uncertain scenarios. f is the objective function to be minimized. $g_j(j = 1, \ldots, p)$ represent a series of constraint functions and p is the number of constraints. Multiple repeated function evaluations of a solution usually get different function values under different uncertain scenarios. Thus, it is difficult to definitely determine the quality of two solutions, which affects the ability of algorithms to find the optimum.

---

[2]The uncertain optimization problem covers both single objective ones and multi-objective ones.

When the number of objectives are multiple, these problems are referred to as uncertain MOPs (UMOPs). A UMOP can be stated as:

$$\text{minimize} \quad \mathbf{F}(\mathbf{x}, \xi) = (f_1(\mathbf{x}, \xi), f_2(\mathbf{x}, \xi), \dots, f_m(\mathbf{x}, \xi)),$$
$$\text{subject to} \quad \mathbf{x} = (x_1, x_2, \dots, x_n) \in \Omega. \tag{1.8}$$

Uncertainties can be in any format such as random, fuzzy, stochastic, rough, and so on. Uncertainties come in different forms, such as stochastic, fuzzy, rough, and compound ones. Uncertainties existing in practice can be categorized into four classes: (1) noise, (2) robustness, (3) fitness approximation, and (4) dynamic fitness functions [Jin and Branke (2005)]. For the last two types of uncertainties, there are specialized techniques to deal with them. Different from deterministic optimization problems, there are no widely accepted test problems for UOPs up to now. In noisy UOPs, noise with different strength levels is implemented as an additive perturbation on well-known deterministic benchmark problems. The majority of studies on robust UOPs concentrate on specific optimization problems, such as job shop scheduling and design optimization. There is no systematic research on the benchmarks of robust UOPs. Besides, the research on the evaluation of optimal solutions of UOPs is still lacking.

1.3 Brief Introduction to Multi-objective Evolutionary Algorithms

The EA is a stochastic method that mimics the natural evolutionary process. Since the EA is a population-based method, it can approximate the whole PF (or PS) at a single run. Rosenberg firstly mentioned the possibility of applying a genetic search to solve MOPs in his doctoral dissertation in 1967. Until 1984, Schaffer extended the genetic algorithm (GA) and put forward an MOEA, namely the vector evaluated genetic algorithm (VEGA) [Schaffer (1985)]. VEGA is essentially a weighted sum method and cannot make a trade-off according to the characteristic of each objective. It can only find the extreme points that are located in the rightmost area when solving MOPs [Schaffer (1984)]. Very often, an approximation of the PF with a manageable number of points and even distribution along the true PF is needed and presented to decision makers for the purpose of supporting their decision-making. Recently, various MOEAs have been widely accepted as major approaches for approximating the true PF [Zhou et al. (2011)]. Based on their selection strategies, these algorithms can be categorized into three classes: (1) Pareto dominance-based

[Deb *et al.* (2002a); Deb and Jain (2014); Fonseca and Fleming (1993); Horn *et al.* (1994); Zitzler *et al.* (2001)]; (2) performance indicator-based [Bader and Zitzler (2011); Zitzler and Kunzli (2004)]; and (3) decomposition-based [Chen *et al.* (2017); Moubayed *et al.* (2014); Qi *et al.* (2014); Zhang and Li (2007)].

1.3.1 *Pareto Dominance-based MOEAs*

The Pareto dominance relationship is the key point of solving MOPs, and thus the Pareto dominance-based MOEA is the most direct way for MOPs. Since the Pareto dominance relationship is a partial order, the dominance relationship is first adopted to rank a set of solutions. Following that, a certain diversity measure is utilized to distinguish solutions in the same Pareto level, which leads to a set of non-dominated solutions widely spread over the PF. By making use of the concept of Pareto dominance relationship, Fonseca and Fleming (1993) introduced a multi-objective genetic algorithm (MOGA). MOGA ranks each individual in a population, adopts the rank-tree based fitness function, and thus associates each individual with the number of solutions which the solution dominates, and the solution is dominated. Besides, the niche scheme is used to obtain a set of uniformly distributed non-dominated solutions. The main drawback of MOGA is that the niche scheme is related with the objective function. Therefore, two different individuals with the same objective function values cannot appear in a population at the same time. Besides, MOGA is simple and easy to implement.

Horn and Nafpliotis put forward a niched-Pareto genetic algorithm (NPGA) [Horn *et al.* (1994)] in 1993. NPGA at first randomly selects two solutions and a comparison set from the population. If one of the two individuals is not dominated by the comparison set, then the individual will be passed to the next generation; otherwise one of them will be selected according to a niching mechanism. NPGA shows an efficient and good performance, but it has difficulty in determining and adjusting the parameter of the niching scheme. Erickson proposed the NPGA2 (2001), which adopts the Pareto ranking scheme, tournament selection and an improved sharing function proposed by Oei to compute the parameter of the niching scheme. However, the dependency of NPGA2 on problem-specific knowledge means that it is not universally used.

Based on the multi-layer classification of individuals, Srinivas and Deb presented a new non-dominated based MOEA, i.e., non-dominated sorting genetic algorithm (NSGA). NSGA classifies a population into different non-dominated fronts and calculates the sharing function values of individuals located on the same front to obtain good diversity of a population. NSGA is good to search the PF, can achieve good population diversity, and allows for the existence of multiple equivalent individuals. However, the convergence speed of NSGA is slow, and obtained solutions show inferior performance than MOGA. To be specific, NSGA depends on sharing parameters and lacks an elitism strategy. Given this, Deb and his research group put forward a non-dominated sorting genetic algorithm with an elitism strategy (NSGA-II) [Deb *et al.* (2002a)]. Firstly, NSGA-II utilizes the fast non-dominated sorting procedure that has a low computation complexity. Then, NSGA-II maintains the parent population and the offspring population at the same time and adopts the elitism mechanism to select best solutions. The disadvantage of NSGA-II is that it is difficult to find isolated points.

Zitzler and Thiele (1999) proposed the strength Pareto evolutionary algorithm (SPEA), which maintains an archive to store non-dominated solutions during the evolutionary process and updates it iteratively. The fitness function value of each solution in SPEA is defined as the number of solutions that the current one dominates. Besides, the clustering method is used to maintain the diversity of a population. Later in 2001, Zitzler *et al.* (2001) proposed an improved version named SPEA2, which utilizes a new fitness function and maintains the population diversity by estimating the density of neighboring solutions. Finally, SPEA2 shows a large improvement compared with SPEA.

1.3.2   *Performance Indicator-based MOEAs*

As we mentioned before, there are various performance metrics that measure the diversity and convergence of a set of non-dominated solutions. Thus, these metrics can be used to guide the search process. Recently, performance indicator-based MOEAs have become a hot research topic in the field of multi-objective optimization. Zitzler first put forward a general indicator-based evolutionary algorithm (IBEA) [Zitzler and Kunzli (2004)]. Any performance indicators can be used to compare solutions and there is no need for an extra diversity maintenance scheme. The *HV*, which can measure both convergence and diversity, is currently the only known metric that is compatible with the Pareto-dominance relationship. It can be proved that maximizing this metric can theoretically guarantee the convergence of the population to the true PF. However, due to the high computation complexity of the *HV*, research has only started on it in 2003. Knowles and Corne first proposed to integrate the *HV* into an evolutionary process and introduced a *HV*-based bounded archive population for the storage of non-dominated solutions. Huband *et al.* (2003) adopted the selection strategy in SPEA and replaced the density measurement with a *HV*-based performance indicator.

When the number of objectives increases, the computational complexity for computing the *HV* increases exponentially. With an aim of reducing computational burden, researchers proposed to estimate the *HV* with a Monte Carlo simulation, which led to the hypervolume estimation algorithm (HypE) [Bader and Zitzler (2011)]. In addition to employing a single metric, Two_Arch [Praditwong and Yao (2006)] first proposed to use two archives that measure convergence and diversity to guide the solution selection. Besides, Wang *et al.* (2015) extended Two_Arch to Two_Arch2 for the usage of solving MaOPs.

1.3.3 *Decomposition-based MOEAs*

Decomposition-based MOEAs gained much attention due to their good properties and became the main approach for optimizing MOPs. Although the concept of the decomposition strategy has been applied to some MOPs [Hughes (2003); Jin *et al.* (2001)], it became famous after the proposition of MOEA/D [Zhang and Li (2007)] in 2007. The main point of the decomposition-based MOEAs is to decompose a complicated MOP into a series of simple subproblems and optimize them in a collaborative way to obtain a set of non-dominated solutions with good convergence and diversity. Decomposition-based MOEAs first attempt to combine classical optimization algorithms with EAs. MOEA/D converts an MOP into a series of scalar subproblems by using aggregation functions (e.g., weighted sum, Tchebycheff, PBI, and so on) and a set of uniformly distributed weight vectors. The concept of a neighbor is defined based on the Euclidean distance among weight vectors. These subproblems are optimized simultaneously by using information from neighboring subproblems, which lead to a low computational complexity. Besides, the diversity of a population is guaranteed by a set of uniformly distributed weight vectors. Recently, researchers have made some improvements on MOEA/D, including the design and adjustment of weight vectors, the integration of new search engine into MOEA/D, the proposal of new aggregation functions, and applying MOEA/D and its variants to different types of practical problems.

MOEA/D adopts a simplex grid-based method [Das and Dennis (1998)] to generate a set of weight vectors whose size increases with the number of objectives and cannot be an arbitrary set. However, it cannot work as well when the target MOP has a complex PF (i.e., discontinuous PF or a PF with a sharp peak or low tail). With the goal of overcoming this drawback, several variants of MOEA/D are presented. Among them, an improved variant of MOEA/D with adaptive weight adjustment (MOEA/D-AWA) [Qi *et al.* (2014)] is put forward. To be specific, the geometric relationship between the weight vectors and the optimal solutions under the Tchebycheff decomposition scheme is analyzed first. Then, a new weight vector initialization method and an adaptive weight vector adjustment strategy are introduced in MOEA/D-AWA. The weights are adjusted periodically so that the weights of the subproblems can be redistributed adaptively to obtain better uniformity of solutions. Meanwhile, computing efforts devoted to subproblems with duplicate optimal solutions can be saved. Moreover, an external elite population is introduced to help add new subproblems into real sparse regions rather than the pseudo sparse regions of the complex PF, i.e., discontinuous regions of the PF. Giagkiozis *et al.* (2013) proposed a generalized decomposition mechanism that can give the optimal setting of weight vectors by optimizing certain scalarizing functions when the referenced PF is given. Li *et al.* (2015b) put forward a two-layer weight vectors generation procedure that is able to overcome the large size of weight vectors induced by the simplex grid-based method. Up to now, it has been adopted by various decomposition-based MOEAs [Deb and Jain (2014); Li *et al.* (2015b)].

In the original version of MOEA/D, all the subproblems are treated equally, each of them receives about the same amount of computational effort. These subproblems, however, may have different computational difficulties, therefore, it is very reasonable to assign different amounts of computational effort to different problems. Zhang *et al.* (2009a) proposed an MOEA/D with dynamical resource allocation (MOEA/D-DRA). In MOEA/D-DRA, a utility value is defined and computed for each subproblem. Computational efforts are distributed to these subproblems based on their utilities. Zhou and Zhang (2015) generalized the dynamical resource allocation in MOEA/D-DRA to a generalized resource allocation (GRA) strategy and embedded it into MOEA/D. MOEA/D-GRA assigns each subproblem with an improvable probability vector based on which allocates the complexity resource for each subproblem.

The simulated binary crossover and the polynomial mutation are utilized to generate new solutions in the original version of MOEA/D. Recently, there has been research on integrating MOEA/D with other heuristic methods to further improve its performance. Zhou and colleagues [Zhou *et al.* (2005)] and Zapotecas-Martínez *et al.* (2015b) adopt the differential evolutionary method and geometrical differential evolutionary algorithm in the framework of MOEA/D, respectively. Furthermore, researchers utilize ant colony optimization [Ke *et al.* (2013)], particle optimization [Moubayed *et al.* (2014)], simulated annealing optimization [Li and Landa-Silva (2011)], adaptive operators [Gonalves *et al.* (2015)], probabilistic models [Zapotecas-Martínez *et al.* (2015a)], and so on, to enhance the performance of MOEA/D.

1.3.4 *MOEAs for Many-objective Optimization Problems*

As mentioned in Section 1.1, MaOPs will induce difficulties to MOEAs that are originally designed for MOPs. In general, there are different ways that have been proposed to overcome the difficulties encountered in evolutionary many-objective optimization. In general, there are five viable ways to alleviate the challenges posed in evolutionary many-objective optimization.

Firstly, for MaOPs that have redundancy among objectives [Brockhoff and Zitzler (2006)], objective reduction [Coello and Chakraborty (2008)] can be an effective approach to convert an MaOP to an MOP with only a few objectives so that the existing MOEAs can be used. It considers employing dimensionality reduction techniques to identify the embedded PF in the ambient objective space. The algorithm in [Brockhoff and Zitzler (2006)] adopts a metric to detect the changes of the dominance structure in the population for objective reduction. Feature selection is used for objective reduction in [Coello and Chakraborty (2008)]. Some dimension reduction techniques in machine learning, such as principal component analysis (PCA) and maximum variance unfolding (MVU) are also adopted in objective reduction [Deb and Saxena (2005); Saxena and Deb (2007); Saxena *et al.* (2013)]. The Pareto corner search evolutionary algorithm (PCSEA) [Singh *et al.* (2011)] is an off-line objective reduction method. Recent studies have demonstrated that some objective reduction-based techniques are vulnerable when the task involves finding a high-dimensional PF [Deb and Saxena (2006)]. However, such techniques are only applicable to problems having a moderate number of conflicting objectives.

In some specific applications, decision makers (DMs) only care about a small region of the entire PF [Cvetkovic and Parmee (2002); Thiele *et al.* (2014)]. Inspired by the success of this idea for the bi-/tri-objective case, researchers have extended their experiments to assess their preference-based approaches for the many-objective case. Hence, preference-based approaches can help decision makers to find parts of PFs [Cvetkovic and Parmee (2002)]. It is the multi-criteria decision-making-based methodology and concerns of finding a preferred subset of solutions from the whole PS. Therefore, the aforementioned difficulties can be alleviated due to the shrunk search space. Several works have been proposed in the literature in this area, such as PBEA [Thiele *et al.* (2014)], *r*-NSGA-II [Said *et al.* (2010)], R-NSGA-II [Deb *et al.* (2006)] and TKRNSGA-II [Bechikh *et al.* (2011)]. However, as noted in [Bechikh (2013)], the preference integration mechanism should not only preserve the order generated by the Pareto dominance but also effectively work whatever the DM preference information is; otherwise inferior results could be obtained due to premature convergence, convergence degradation, and so on.

As to the non-reduced MaOP, most Pareto-based MOEAs experience difficulties in finding the whole PF due to the above-mentioned difficulties. In order to alleviate these challenges, the first and most straightforward one is the development of new dominance relations, which can refine or enhance the Pareto dominance relation to increasing the selection pressure towards the PF. So far, a large amount of studies along this direction have been reported, such as dominance area control [Branke *et al.* (2001)], *k*-optimality [Farina and Amato (2004)], preference order ranking [Pierro *et al.* (2007)], subspace dominance comparison [Jaimes *et al.* (2011)], ε-dominance [Tanaka (2009)], and so on. Although these modified dominance relationships have been shown to be able to improve the convergence of MOEAs for solving MaOPs, they may also cause the population to converge to a subregion of the PF. The strengths and weaknesses of different preference relations on MaOPs have been experimentally investigated to a certain degree by several similar studies [Coello (2009)].

As a result of the increase of the objective space, the conflict between the convergence and diversity requirements is gradually aggravated [Adra and Fleming (2011); Purshouse and Fleming (2007)]. The diversity-based secondary selection schemes play a crucial role in determining the survival of individuals. Most of the diversity maintenance techniques (e.g., niche, crowding distance, clustering and kth nearest distance) not only reduce the selection pressure toward the PF, but also hinder the evolutionary search to some extent because they favor dominance resistant solutions [Purshouse and Fleming (2007)]. The consideration is that since Pareto dominance is unable to exert sufficient selection pressure towards the PF in a high-dimensional objective space, then the selection should be almost solely based on the diversity, which is generally regarded as the secondary selection operator in MOEAs. However, most existing diversity maintenance criteria, e.g., crowding distance, prefer dominance resistant solutions, which would cause the search to be biased towards solutions with poor proximity to the global PF, although these solutions may present good diversity over the objective space. Thus, the diversity preservation in the many-objective scenarios requires some care to avoid or weaken this phenomenon. Apparently, an alternative avenue to enhance the ability of dominance-based MOEAs for MaOPs is either to improve the diversity maintenance mechanism or to replace it with other selection criteria. Compared to the first type of research, the work along this direction [Deb and Jain (2014); Li *et al.* (2014)] has received much less attention.

Unlike Pareto-dominance-based MOEAs, another two classes of MOEAs, indicator- and decomposition-based MOEAs, have not been criticized much on the selection pressure produced in a high-dimensional objective space. Indicator-based evolutionary algorithms (IBEAs) [Zitzler and Kunzli (2004)] adopt a single indicator, which accounts for both convergence and distribution performances of a solution set, such as hypervolume [Zitzler *et al.* (2003)], to guide the selection process. Therefore, solutions can be selected one by one based on their influence on the performance indicator. In this class, an MaOP is converted into the problem of optimizing an indicator by evaluating the solutions using a performance metric. However, they would encounter their own difficulties when handling MaOPs. The exponentially increased computation cost of hypervolume calculation, as mentioned in the fourth challenge, severely impedes further developments of indicator-based MOEAs for many-objective optimization. Although some efforts have been made to remedy the computational issue, it is still far from widely applicable in practice. To address this issue, one of the alternative strategies is to estimate the rankings of solutions induced by the *HV* indicator without computing the exact indicator values [Bader and Zitzler (2011)]. Another strategy is to replace the *HV* with other indicators that are much less computationally expensive but also have good theoretical properties. Recently, R2 and the additive approximation were also proposed to further enhance the computational efficiency of IBEAs for solving MaOPs [Hernandez Gomez and Coello Coello (2013)].

In many-objective optimization, it is indeed very difficult for an MOEA to emphasize both convergence and diversity within a single population or archive. Motivated by this, another type of approach separately maintains two archives for convergence and diversity during the evolutionary search, and at least one of the two archives still uses the Pareto dominance. The related work can be referred to in [Praditwong and Yao (2006); Wang *et al.* (2015)]. Intuitively, it is likely to consider convergence and distribution of a solution set separately, e.g., by simultaneously measuring the distance of the solutions to the PF, and maintaining a sufficient distance between each other. Motivated by this idea, a many-objective evolutionary algorithm using a one-by-one selection strategy, also known as 1by1EA, is proposed in [Liu *et al.* (2017)]. In 1by1EA, solutions are selected according to a convergence indicator and a distribution indicator. The former measures the distance between a solution and the PF, while the latter measures its distance to each other. In the environmental selection, offspring individuals are selected one by one based on a computationally efficient convergence indicator to increase the selection pressure toward the PF. In 1by1EA, once an individual is selected, its neighbors are de-emphasized using a niche technique to guarantee the diversity of the population, in which the similarity between individuals is evaluated by means of a distribution indicator. In addition, different methods for calculating the convergence indicator are examined and an angle-based similarity measure is adopted for effective evaluations of the distribution of solutions in the high-dimensional objective space.

Recently, decomposition has emerged as a promising approach to tackle many-objective optimization. Some decomposition-based algorithms have been found to be successful in both multi and many-objective problems. The decomposition-based method decomposes an MOP, in question, into a set of subproblems and optimizes them in a collaborative manner. Note that the decomposition concept is so general that either aggregation functions or simpler MOPs [Liu *et al.* (2014)] can be used to form subproblems. The MOEA based on decomposition (MOEA/D) [Zhang and Li (2007)] and multiple single-objective Pareto sampling (MSOPS) [Hughes (2003)] are two most typical decomposition-based MOEAs. Among them, MOEA/D [Zhang and Li (2007)] is the most well-known. In MOEA/D, an MOP is decomposed into a number of subproblems and each population member is assigned a weight vector and optimizes its related subproblem based on a scalarizing function. Each solution is associated with a subproblem, and each subproblem is optimized by using information from its neighborhoods. During the past few years, MOEA/D has spawned a large amount of research work, e.g., introducing adaptive mechanism in reproduction [Chen *et al.* (2009a); Gonalves *et al.* (2015); Huang and Li (2010)], hybridizing with local search [Sindhya *et al.* (2011)], and incorporating stable matching in selection.

Recently, some studies in the literature adopted the fusion of Pareto dominance and decomposition-based approaches [Deb and Jain (2014); Li *et al.* (2015b)]. This observation highly motivates us to combine the Pareto and non-Pareto selection criteria so that we can exploit the merits from both approaches. Li *et al.* (2015b) introduced MOEA/DD that combines the Pareto dominance and decomposition. It uses a non-dominated sorting to sort the population into different non-domination levels. In order to maintain the diversity of the population, a solution in the last considered non-domination level will survive in case it is associated with an isolated subregion. Unlike MOEA/D, the NSGA-III [Deb and Jain (2014)] decomposes an MaOP based on a set of well-distributed reference points that are created using Das and Dennis' systematic approach. Recall that the classical NSGA-II uses a niching strategy to select individuals from the last acceptable front, which is situated in the least crowded regions. This selection mechanism was modified in NSGA-III. Instead of explicitly measuring the crowding distance, a reference point-based niche preservation operation is performed to choose the solutions in the last acceptable level. Very recently, Yuan *et al.* (2016) proposed a new framework based on NSGA-II for many-objective optimization, called ensemble fitness ranking (EFR), which is more general than MSOPS.

1.4 Brief Introduction to the Uncertain Optimization

During the past decades, handling uncertainties has been a hot topic, and there are several ways to tackle uncertainties in the literature. To our knowledge, existing approaches for solving uncertain problems can be grouped into the following three categories.

1.4.1   *Converting*

It means converting uncertain functions into their deterministic counterparts. The uncertainty theory founded by Liu (2009) is a branch of mathematics based on normality, duality, subadditivity and product axioms. It is a common way to transform uncertain functions into deterministic ones. Based on the uncertainty theory, the chance-constrained programming and the dependent-chance programming have been proposed and applied to deal with practical problems [Wang *et al.* (2014)]. Additionally, mean and robust approaches are commonly used strategies for handling uncertainties. For example, an evolutionary algorithm based on a mean operator for multiobjective flow-shop scheduling problems with uncertainties was suggested in [Liefooghe *et al.* (2007)]. Actually, the mean operator is inappropriate when the distributions of uncertain functions are far from normal. A robust approach was adopted in [Li *et al.* (2016a)] and [Xiong *et al.* (2017)]. The robust approach is a pessimistic approach with a high risk aversion, and it may be too conservative and restricting. In [Rockafellar and Uryasev (2000)], a percentile risk measure, which is called conditional value-at-risk (CVaR) was adopted in stochastic programming with poorly defined distributions. It took a single-objective two-stage weapon-target assignment (WTA) problem as an illustrative example.

1.4.2   *Designing*

It means that we can design different uncertainty-tolerant algorithms or embed uncertainty-handling components in existing algorithms. Although meta-heuristics are known to be inherently robust to low-level uncertainties due to their distributed nature and nonreliance on gradient information, uncertainty-handling components are still needed to reduce detrimental effects of high-level uncertainties. For instance, Buche *et al.* (2002) proposed the noise tolerant strength Pareto evolutionary algorithm (NTSPEA) with an improved robust performance against noise. Teich (2001) and Hughes (2001) both introduced a probabilistic Pareto ranking scheme to account for the presence of uncertainties. Using NSGA-II as the baseline algorithm, Singh and Minsker (2008) extended the nonsampling method and the probabilistic selection scheme [Hughes (2001)] to solve a noisy groundwater remediation design problem. Fieldsend and Everson (2015) proposed a novel algorithm, named the rolling tide evolutionary algorithm (RTEA), to cope with UMOPs.

1.4.3 *Using Multi-objective Approaches*

It means that we can define an additional objective to describe the uncertainty. For example, the mean-variance model [Huang (2007)], the mean-semi-variance model [Huang (2008)], the mean-variance-skewness model [Li *et al.* (2010b)] and the mean-entropy model [Sulieman *et al.* (2010)] are commonly used approaches. However, these types of methods will increase the number of objectives and thus increase the complexity of the problem to be handled.

Chapter 2

Decomposition-based Multi-objective Evolutionary Algorithm with the ε-Constraint Framework

Decomposition is an efficient and prevailing strategy for solving multi-objective optimization problems (MOPs). Its success has been witnessed by the multi-objective evolutionary algorithm MOEA/D and its variants. In decomposition-based methods, an MOP is decomposed into a number of scalar subproblems by using various scalarizing functions. Most decomposition schemes adopt the weighting method to construct scalarizing functions. In this chapter, another classical generation method in the field of mathematical programming, that is the ε-constraint method, is adopted for the multi-objective optimization. It selects one of the objectives as the main objective and converts other objectives into constraints. We incorporate the ε-constraint method into the decomposition strategy and propose a new decomposition-based multi-objective evolutionary algorithm with the ε-constraint framework (DMOEA-εC). It decomposes an MOP into a series of scalar-constrained optimization subproblems by assigning each subproblem with an upper bound vector. These subproblems are optimized simultaneously by using information from neighboring subproblems. Besides, a main objective alternation strategy, a solution-to-subproblem matching procedure, and a subproblem-to-solution matching procedure are proposed to strike a balance between convergence and diversity.

2.1 Related Algorithms

The most popular decomposition-based MOEA is MOEA/D [Zhang and Li (2007)]. It decomposes an MOP into a number of scalar optimization subproblems and then applies evolutionary algorithms (EAs) to optimize these subproblems in a collaborative manner. Each subproblem is optimized by using information from its neighboring subproblems, which makes

MOEA/D with relative low computational complexity. Neighborhood relations between these subproblems are defined based on Euclidean distances between their weight vectors. The diversity of the population is implicitly maintained by specifying a good spread of the weight vectors in the objective space. In recent years, MOEA/D has attracted increasing research interests and many follow-up studies have been published. In [Chen *et al.* (2009a); Huang and Li (2010); Lai (2009); Li and Landa-Silva (2011); Li *et al.* (2010a); Moubayed *et al.* (2010); Shim *et al.* (2012); Sindhya *et al.* (2011)], authors combine MOEA/D with other metaheuristics. Jain and Deb (2014b) and Ishibuchi *et al.* (2009, 2010) research on different scalarizing approaches. Besides, some studies address the problem of adjusting weight vectors to make optimal solutions uniformly distributed along the PF [Gu and Liu (2010); Li and Landa-Silva (2011); Qi *et al.* (2014)] and extending applications from benchmark problems to real-world problems [Laili *et al.* (2016); Li *et al.* (2010a, 2015a); Lu *et al.* (2016); Mei *et al.* (2011); Wang *et al.* (2013); Zhang *et al.* (2010)].

In general, the weighting method and the ε-constraint method are two basic generation methods [Miettinen (1999)]. They are often used as elements of more developed methods. MOEA/D [Zhang and Li (2007)] takes inspiration from the weighting method. The ε-constraint method selects one of the objectives as the main objective while transforming the other non-main objectives to constraints and associating each non-main objective with an upper bound coefficient. Mavrotas (2009) first proposed an augmented ε-constraint method (AUGMECON) that divides the range of each non-main objective function into a fixed number of equal intervals by using the ideal and nadir points obtained from the payoff table and uses these grid points as upper bound vectors. Besides, in order to avoid the production of weakly Pareto optimal solutions, AUGMECON transforms inequality constraints to equality ones by explicitly incorporating slack variables. In the literature, several versions of the ε-constraint method have been put forward. Mavrotas and Florios (2013) presented the AUGMECON2, which is an improvement of AUGMECON. It introduces a concept of the bypass coefficient that indicates how many consecutive iterations can be bypassed. Grandinetti *et al.* (2013) proposed an approximate ε-constraint method to solve a multi-objective job scheduling problem. The method defines a finite sequence of ε-constraint problems through a progressive reduction of the values of upper bound coefficients. Zhang and Reimann (2014) put forward a simple augmented ε-constraint method (SAUGMECON), which is a variant of AUGMECON2. The innovative

mechanisms of SAUGMECON include an extension to the acceleration algorithm with an early exit and an addition of an acceleration algorithm with
bouncing steps.

2.2 Formulation of MOPs under the ε-Constraint Framework

In the ε-constraint method firstly introduced by Haimes *et al.* (1971), one
of the objectives is selected as the main objective function to be optimized
and all the other non-main objectives are converted into constraints by
giving an upper bound coefficient to each of them. The ε-constraint problem
corresponding to the MOP (1.1) is formulated as follows:

$$\text{minimize} \quad f_{main} = f_s(\mathbf{x}) + \rho \sum_{i=1}^{m} f_i(\mathbf{x})$$

$$\text{subject to} \quad \begin{cases} \dfrac{f_i(\mathbf{x}) - z_i^*}{z_i^{nad} - z_i^*} \le \varepsilon_i, \forall i \in \{1, 2, \cdots, m\} \backslash \{s\} \\ \mathbf{x} = (x_1, x_2, \ldots, x_n) \in \Omega \end{cases} \quad (2.1)$$

where $0 \le \varepsilon = (\varepsilon_1, \ldots, \varepsilon_{s-1}, \varepsilon_{s+1}, \ldots, \varepsilon_m) \le 1$ is the upper bound vector. The main objective index s is randomly selected from $\{1, 2, \ldots, m\}$
or predefined by decision makers. $\rho > 0$ is a very small positive num-
ber. $\mathbf{z}^* = (z_1^*, \ldots, z_m^*)$ and $\mathbf{z}^{nad} = (z_1^{nad}, \ldots, z_m^{nad})$ are the ideal point
and the nadir point, respectively. The exact definitions will be given in
the following. An example illustration of different upper bounds for the
ε-constraint method is shown in Fig. 2.1. In Fig. 2.1, different upper bounds
(i.e., different ε values) for the objective function f_2 are given, while the
objective function f_1 is selected as the main objective function to be minimized. The Pareto optimal solutions corresponding to the four upper bound
vectors are shown by the black points.

2.3 Related Theoretical Results

Theorem 2.1. *For any given upper bound vector* $\varepsilon = (\varepsilon_1, \ldots, \varepsilon_{s-1}, \varepsilon_{s+1}, \ldots, \varepsilon_m)$, *the optimal solution of the problem* (2.1) *is the Pareto optimal for the original MOP* (1.1).

Proof. Let $\mathbf{x}^*$ be an optimal solution of the problem (2.1) for some given
upper bound vector $0 \le \varepsilon = (\varepsilon_1, \ldots, \varepsilon_{s-1}, \varepsilon_{s+1}, \ldots, \varepsilon_m) \le 1$, which
means that $f_s(\mathbf{x}^*) + \rho \sum_{i=1}^{m} f_i(\mathbf{x}^*) \le f_s(\mathbf{x}) + \rho \sum_{i=1}^{m} f_i(\mathbf{x})$ for all $\mathbf{x} \in \Omega$

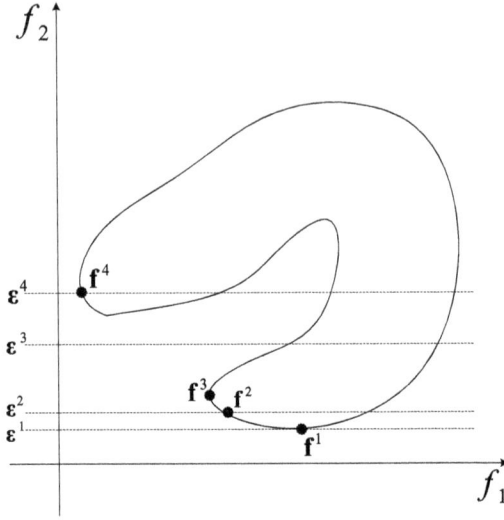

Fig. 2.1   Different upper bounds for the ε-constraint method.

when $\frac{f_i(\mathbf{x}^*)-z_i^*}{z_i^{nad}-z_i^*} \leq \varepsilon_i$, equivalently, $f_i(\mathbf{x}^*) \leq \varepsilon_i(z_i^{nad} - z_i^*) + z_i^*$, for every $i = 1, \ldots, m, i \neq s$ and $\mathbf{x} \in \Omega$. Let us assume that $\mathbf{x}^*$ is not a Pareto optimal solution for the original problem (1.1). In this case, there exists a vector $\mathbf{x_0} \in \Omega$ such that $f_i(\mathbf{x_0}) \leq f_i(\mathbf{x}^*)$ for all $i = 1, \ldots, m$, and the inequality is strict for at least one index j. It is clear that $\mathbf{x_0}$ is a feasible solution of problem (2.1) and the inequality $f_s(\mathbf{x_0}) + \rho \sum\limits_{i=1}^{m} f_i(\mathbf{x_0}) < f_s(\mathbf{x}^*) + \rho \sum_{i=1}^{m} f_i(\mathbf{x}^*)$ holds. Here there is a contradiction with the fact that $\mathbf{x}^*$ is an optimal solution of the problem (2.1).

The conclusion drawn from the above is that if $\mathbf{x}^*$ is an optimal solution of the problem (2.1) for some given upper bound vector $0 \leq \varepsilon = (\varepsilon_1, \ldots, \varepsilon_{s-1}, \varepsilon_{s+1}, \ldots, \varepsilon_m) \leq 1$, then $\mathbf{x}^*$ must be a Pareto optimal solution for the original MOP (1.1). $\square$

Theorem 2.2. *Let $\mathbf{x}^*$ be a proper Pareto optimal solution to the problem (1.1), then there exists an upper bound vector $\varepsilon = (\varepsilon_1, \ldots, \varepsilon_{s-1}, \varepsilon_{s+1}, \ldots, \varepsilon_m)$ with $0 \leq \varepsilon_i \leq 1, \forall i \in \{1, 2, \ldots, m\} \backslash \{s\}$ and a real number $0 < \rho < \frac{1}{(m-1)M-1}$ such that $\mathbf{x}^*$ is an optimal solution to the problem (2.1).*

Proof. Construct an upper bound vector $\varepsilon = (\varepsilon_1, \ldots, \varepsilon_{s-1}, \varepsilon_{s+1}, \ldots, \varepsilon_m)$ in the following way: $\varepsilon_i = \frac{f_i(\mathbf{x}^*)-z_i^*}{z_i^{nad}-z_i^*}, i = 1, \ldots, m, i \neq s.$

Obviously, the feasible region of (2.1), denoted by $R_f = \{\mathbf{x}|\mathbf{x} \in \Omega \wedge \frac{f_i(\mathbf{x})-z_i^*}{z_i^{nar}-z_i^*} \le \varepsilon_i, i = 1, \dots, m, i \ne s\}$, is nonempty since $\mathbf{x}^* \in R_f$. Then, for any $\mathbf{x} \in R_f$, from $\varepsilon_i = \frac{f_i(\mathbf{x}^*)-z_i^*}{z_i^{nad}-z_i^*}$, it can be derived that $f_i(\mathbf{x}) \le f_i(\mathbf{x}^*)$ always holds for each $i \in 1, \dots, m(i \ne s)$. In addition, if $f_s(\mathbf{x}) < f_s(\mathbf{x}^*)$, then $\mathbf{x}^*$ will be dominated by $\mathbf{x}$, which contradicts the fact that $\mathbf{x}^*$ is Pareto optimal. Therefore, $f_s(\mathbf{x}^*) \le f_s(\mathbf{x})$ holds. Next, we complete the proof from two cases:

Case 1. $f_s(\mathbf{x}^*) = f_s(\mathbf{x})$
In this case, $f_s(\mathbf{x}^*) = f_s(\mathbf{x})$ holds for each $i \in 1, \dots, m(i \ne s)$, otherwise $\mathbf{x}^*$ will be dominated by $\mathbf{x}$, which contradicts the fact that $\mathbf{x}^*$ is Pareto optimal. Therefore, $f_s(\mathbf{x}^*)+\rho\sum_{i=1}^{m} f_i(\mathbf{x}^*) = f_s(\mathbf{x})+\rho\sum_{i=1}^{m} f_i(\mathbf{x})$ holds for any $\mathbf{x} \in R_f$ satisfying $f_s(\mathbf{x}^*) = f_s(\mathbf{x})$.

Case 2. $f_s(\mathbf{x}^*) < f_s(\mathbf{x})$
In this case, we have the following two subcases:

Case 2.1. $\sum_{i=1}^{m} f_i(\mathbf{x}^*) \le \sum_{i=1}^{m} f_i(\mathbf{x})$: we can obtain $f_s(\mathbf{x}^*)+\rho\sum_{i=1}^{m} f_i(\mathbf{x}^*) < f_s(\mathbf{x}) + \rho\sum_{i=1}^{m} f_i(\mathbf{x})$ for any $\mathbf{x} \in R_f$ satisfying $f_s(\mathbf{x}^*) < f_s(\mathbf{x})$.

Case 2.2. $\sum_{i=1}^{m} f_i(\mathbf{x}^*) > \sum_{i=1}^{m} f_i(\mathbf{x})$. Combined with $f_i(\mathbf{x}) \le f_i(\mathbf{x}^*)$ for each $i \in 1, \dots, m(i \ne s)$, then there must exist at least one index $j \ne s$ such that $f_j(\mathbf{x}) < f_j(\mathbf{x}^*)$. We take f_k satisfying $f_k(\mathbf{x}^*)-f_k(\mathbf{x}) = \max_j\{f_j(\mathbf{x}^*)-f_j(\mathbf{x})\}$.

Since $\mathbf{x}^*$ is properly optimal, we have $\frac{f_k(\mathbf{x}^*)-f_k(\mathbf{x})}{f_s(\mathbf{x})-f_s(\mathbf{x}^*)} \le M$.
Then,

$$f_s(\mathbf{x}) + \rho\sum_{i=1}^{m} f_i(\mathbf{x}) - \left(f_s(\mathbf{x}^*) + \rho\sum_{i=1}^{m} f_i(\mathbf{x}^*)\right)$$

$$= (1+\rho)(f_s(\mathbf{x}) - f_s(\mathbf{x}^*)) + \rho \sum_{i=1,i\ne s}^{m} (f_i(\mathbf{x})-f_i(\mathbf{x}^*))$$

$$\ge (1+\rho)(f_s(\mathbf{x}) - f_s(\mathbf{x}^*)) + \rho(m-1)(f_k(\mathbf{x}) - f_k(\mathbf{x}^*))$$

$$\ge (1+\rho)(f_s(\mathbf{x}) - f_s(\mathbf{x}^*)) + \rho(m-1)M(f_s(\mathbf{x}^*) - f_s(\mathbf{x}))$$

$$= \rho\left(\frac{1+\rho}{\rho} - (m-1)\cdot M\right)\cdot(f_s(\mathbf{x}^*) - f_s(\mathbf{x})).$$

Since $0 < \rho < \frac{1}{(m-1)M-1}$ and $f_s(\mathbf{x}^*) < f_s(\mathbf{x})$, we get $f_s(\mathbf{x}) + \rho\sum_{i=1}^{m} f_i(\mathbf{x}) - (f_s(\mathbf{x}^*) + \rho\sum_{i=1}^{m} f_i(\mathbf{x}^*)) > 0$, equivalently, $f_s(\mathbf{x}^*) + \rho\sum_{i=1}^{m} f_i(\mathbf{x}^*) < f_s(\mathbf{x}) + \rho\sum_{i=1}^{m} f_i(\mathbf{x})$ for any $\mathbf{x} \in R_f$ satisfying $f_s(\mathbf{x}^*) < f_s(\mathbf{x})$.

From the above two cases, for any $\mathbf{x} \in R_f$, whether it satisfies $f_s(\mathbf{x}^*) = f_s(\mathbf{x})$ or $f_s(\mathbf{x}^*) < f_s(\mathbf{x})$, we can have the conclusion that $f_s(\mathbf{x}^*) + \rho \sum_{i=1}^m f_i(\mathbf{x}^*) \leq f_s(\mathbf{x}) + \rho \sum_{i=1}^m f_i(\mathbf{x})$, which means $\mathbf{x}^*$ is an optimal solution of the problem (2.1).

Then, it can be concluded that for any proper Pareto optimal solution $\mathbf{x}^*$ of the original MOP (1.1), there exists an upper bound vector $0 \leq \varepsilon = (\varepsilon_1, \ldots, \varepsilon_{s-1}, \varepsilon_{s+1}, \ldots, \varepsilon_m) \leq 1$ and a real number $0 < \rho < \frac{1}{(m-1)M-1}$ is the optimal solution of the problem (2.1). $\qquad\square$

Theorem 2.3. *For any MOP with bounded Pareto front (PF), Pareto optimality is equivalent to proper Pareto optimality. That is to say, if $\mathbf{x}^* \in \Omega$ is a Pareto optimal solution for an MOP with bounded PF, then $\mathbf{x}^* \in \Omega$ is a proper Pareto optimal solution for the same MOP.*

Proof. Take the minimization problem $P0$ as an example for illustration. For each f_i and $\mathbf{x} \in \Omega$ satisfying $f_i(\mathbf{x}) < f_i(\mathbf{x}^*)$, assume that $f_j(\mathbf{x}) \leq f_j(\mathbf{x}^*)$ holds for each $j \in 1, \ldots, m (j \neq i)$ (m is the number of objectives), then $\mathbf{x}^*$ is dominated by $\mathbf{x}$, which contradicts the fact that $\mathbf{x}^*$ is Pareto optimal. Thus for each f_i and $\mathbf{x} \in \Omega$ satisfying $f_i(\mathbf{x}) < f_i(\mathbf{x}^*)$, there exists at least one f_j such that $f_j(\mathbf{x}^*) < f_j(\mathbf{x})$.

Since the PF is bounded, take $M = \dfrac{\max\limits_{i \in S_{\mathbf{x} < f \mathbf{x}^*}} (f_i(\mathbf{x}^*) - f_i(\mathbf{x}))}{\min\limits_{j \in S_{\mathbf{x}^* < f \mathbf{x}}} (f_j(\mathbf{x}) - f_j(\mathbf{x}^*))} > 0$ where $S_{\mathbf{x} < f \mathbf{x}^*} = \{i | f_i(\mathbf{x}) < f_i(\mathbf{x}^*), i = 1, \ldots, m\}$, $S_{\mathbf{x}^* < f \mathbf{x}} = \{j | f_j(\mathbf{x}^*) < f_j(\mathbf{x}), j = 1, \ldots, m\}$. Therefore, we have $\dfrac{f_i(\mathbf{x}^*) - f_i(\mathbf{x})}{f_j(\mathbf{x}) - f_j(\mathbf{x}^*)} \leq M$.

The above discussion implies that there is some real number $M = \dfrac{\max\limits_{i \in S_{\mathbf{x} < f \mathbf{x}^*}} (f_i(\mathbf{x}^*) - f_i(\mathbf{x}))}{\min\limits_{j \in S_{\mathbf{x}^* < f \mathbf{x}}} (f_j(\mathbf{x}) - f_j(\mathbf{x}^*))} > 0$ for each f_i and $\mathbf{x} \in \Omega$ satisfying $f_i(\mathbf{x}) < f_i(\mathbf{x}^*)$, there exists at least one f_j such that $f_j(\mathbf{x}^*) < f_j(\mathbf{x})$ and $\dfrac{f_i(\mathbf{x}^*) - f_i(\mathbf{x})}{f_j(\mathbf{x}) - f_j(\mathbf{x}^*)} \leq M$. Therefore, $\mathbf{x} \in \Omega$ is a properly Pareto optimal solution for the same MOP. $\qquad\square$

Theorem 2.4. *For any MOP with a bounded PF, let $\mathbf{x}^*$ be a Pareto optimal solution to the problem (1.1), then there exists an upper bound vector $\varepsilon = (\varepsilon_1, \ldots, \varepsilon_{s-1}, \varepsilon_{s+1}, \ldots, \varepsilon_m)$ with $0 \leq \varepsilon_i \leq 1, \forall i \in \{1, 2, \cdots, m\} \backslash \{s\}$ such that $\mathbf{x}^*$ is an optimal solution to the problem (2.1).*

Theorem 2.4 is a straightforward extension of Theorem 2.2 with the help of Theorem 2.3.

To sum up, for any given $\varepsilon = (\varepsilon_1, \ldots, \varepsilon_{s-1}, \varepsilon_{s+1}, \ldots, \varepsilon_m)$, the optimal solution of the problem (2.1) is Pareto optimal for the original MOP (1.1);

for each Pareto optimal solution to the problem (1.1), there exists an upper bound vector $\varepsilon = (\varepsilon_1, \ldots, \varepsilon_{s-1}, \varepsilon_{s+1}, \ldots, \varepsilon_m)$ such that the Pareto optimal solution of (1.1) is an optimal solution to the problem (2.1). Due to the fact that the PFs of most practical problems are bounded, the assumption of the bounded PF is reasonable. Thus Theorem 2.4 is applicable for a wide range of problems.

2.4   Framework of DMOEA-εC

2.4.1   *Algorithmic Framework*

DMOEA-εC converts an MOP into N scalar-constrained subproblems and optimizes them simultaneously in a single run. Let $\{\varepsilon^1, \varepsilon^2 \ldots, \varepsilon^N\}$ be a set of evenly spread upper bound vectors, and the neighborhood of upper bound vector ε^i is defined as a set of its several closest upper bound vectors in $\{\varepsilon^1, \varepsilon^2 \ldots, \varepsilon^N\}$. The neighborhood of the ith subproblem consists of all subproblems with the upper bound vectors from the neighborhood of ε^i and is denoted as $B(i)$. When optimizing each subproblem, the feasibility rule [Deb (2000)] is adopted as the constraint handling method. When two solutions are compared, the following criteria are followed:

1) Any feasible solution is preferred to any infeasible solution.
2) Among two feasible solutions, the one having better objective function value is preferred.
3) Among two infeasible solutions, the one having smaller constraint violation is preferred.

The following notations will be used in the description of DMOEA-εC:

- N: The number of upper bound vectors, which is the same as the population size;
- T: Neighborhood size;
- δ: Probability of selecting mate solutions from its neighborhood;
- n_r: Maximum number of replacement;
- IN_m: Iteration interval of alternating the main objective function;
- $DRA\_interval$: Iteration interval of utilizing the dynamic resource allocation strategy;
- S: Maximum size of the external archive population; and
- NFE: Maximum number of function evaluations.

The algorithmic description of DMOEA-εC is presented in **Algorithm 1**. In this algorithm, *rand* is a uniformly randomly distributed value in $[0, 1]$.

2.4.2   *Generation of Upper Bound Vectors* $\{\varepsilon^1, \varepsilon^2 \ldots, \varepsilon^N\}$

A structured set of upper bound vectors is generated by dividing each objective axis with an equal spacing Δ. The process of generating the upper bound vectors is illustrated by using a tri-objective optimization problem ($m = 3$) with a spacing of $\Delta = 1/4(q = 5)$ in Fig. 2.2. The process results in the generation of 25 upper bound vectors.

2.4.3   *Solution-to-Subproblem Matching Procedure*

After the main objective alternation strategy is utilized, a solution which is good for the current subproblem will no longer perform well since the objective function of this subproblem has been changed. Thus a solution-to-subproblem matching procedure is proposed. In this matching procedure, the solution that has the minimum distance value to certain subproblems among N current solutions is matched with this subproblem. The distance value from solution $(\mathbf{x}^i, \mathbf{FV}^i)$ to the subproblem with ε^l is defined as $d_i^l = \sum_{j=1, j \neq s}^{m} |f_j^i - \varepsilon_j^l|$ as described in **Algorithm 2**. This matching procedure will be used after the main objective alternation procedure. It can place the nearest solution to each subproblem at a large degree, which is beneficial to diversity.

2.4.4   *Dynamic Resource Allocation Strategy*

Different subproblems have different computational difficulties, therefore, it is reasonable to assign different amounts of computational effort to them based on their utility values, which are defined in line 3 of **Algorithm 3** [Zhang *et al.* (2009a)]. The 10-tournament selection in line 5 is used to select subproblems that need to be processed in the next generation. It means drawing 10 subproblems from the set of all subproblems randomly, and the subproblem with the best utility value will be selected. This process is repeated $\lfloor \frac{N}{5} \rfloor - m - 1$ times to obtain the indices of the selected subproblems.

2.4.5   *Subproblem-to-Solution Matching Procedure*

When a new solution is generated, it may perform badly for the current subproblem but perform well for another subproblem. In order to avoid

---

Algorithm 1 Framework of DMOEA-εC

Require: An MOP, related parameters.
Ensure: An external archive population EP.
1: Initialize N evenly spread upper bound vectors; randomly initialize the evolving population $\mathbf{P} = \{\mathbf{x}^1, \ldots, \mathbf{x}^N\}$ and set $\mathbf{FV}^i = \mathbf{F}(\mathbf{x}^i)$; extract non-dominted individuals from P and denote the set of them as EP; initialize $\mathbf{z}^*$ and $\mathbf{z}^{\mathrm{nad}}$; set $gen = 0, n = N$.
2: Use the solution-to-subproblem matching procedure (**Algorithm 2**) to match solutions with subproblems.
3: **for** $i = 1$ to N **do**
4:    Set the neighborhood of the ith subproblem $B(i)$.
5:    $\pi^i = 1$.
6: **end for**
7: **while** $n \le NFE$ **do**
8:    **if** gen is a multiple of $DRA\_interval$ **then**
9:        Update the indices of the subproblems I that will be processed in next generation by applying the dynamic resource allocation scheme (**Algorithm 3**).
10:    **end if**
11:    **if** gen is a multiple of IN_m **then**
12:        Alternate the main objective.
13:        Use the solution-to-subproblem matching procedure (**Algorithm 2**) to match solutions with subproblems.
14:    **end if**
15:    **for** $i \in I$ **do**
16:        $P = \begin{cases} B(i), & \text{if } rand < \delta \\ \{1, 2, \ldots, N\}, & otherwise \end{cases}$
17:        Reproduction: Select parent individuals from P randomly and apply certain reproduction operator to generate a new solution $\mathbf{y}$.
18:        $n = n + 1$.
19:        Repair: If $\mathbf{y}$ is infeasible, repair it.
20:        Update the approximated ideal point $\mathbf{z}^*$.
21:        Use the subproblem-to-solution matching procedure (**Algorithm 4**) to find a subproblem k for $\mathbf{y}$.
22:        Compare $\mathbf{y}$ with neighboring solutions of the subproblem k and update these neighboring solutions by using the feasibility rule.
23:        Update the external archive EP and prune it by using the farthest-candidate approach (**Algorithm 5**).
24:        Update the approximated nadir point $\mathbf{z}^{\mathrm{nad}}$.
25:    **end for**
26:    $gen = gen + 1$.
27: **end while**

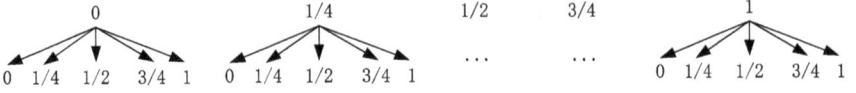

Fig. 2.2   Illustration of generating uniformly spread upper bound vectors ($m = 3, q = 5$).

Algorithm 2 Solution-to-Subproblem Matching

Require: N solutions $(\mathbf{x}^1, \mathbf{FV}^1), \ldots, (\mathbf{x}^N, \mathbf{FV}^N)$ and N subproblems with upper bound vectors $\varepsilon^1, \varepsilon^2 \ldots, \varepsilon^N$.
Ensure: Matched pairs $(\mathbf{x}^k, \mathbf{FV}^k) \sim \varepsilon^l(k, l = 1, \ldots, N)$.
  1: Initialize $S = \{1, 2, \ldots, N\}$.
  2: **while** S is nonempty **do**
  3:     Randomly select an upper bound vector $\varepsilon^l, l \in S$.
  4:     **for** $i = 1$ to N **do**
  5:       $d_i^l = \sum_{j=1, j \neq s}^{m} |f_j^i - \varepsilon_j^l|$.
  6:     **end for**
  7:     $k = argmin_{i=1,\ldots,N}(d_1^l, d_2^l, \ldots, d_N^l)$.
  8:     $\mathbf{FV}^k = \mathbf{Inf}$; $S = S \backslash \{l\}$.
  9: **end while**

Algorithm 3 Dynamic Resource Allocation [Zhang *et al.* (2009a)]

Require: Utility values $\pi^1, \pi^2, \ldots, \pi^N$, old and new function values of each subproblem, denoted as $f_{main}^{old\_i}, f_{main}^{new\_i}$, respectively, for all $i = 1, 2, \ldots, N$.
Ensure: The indices of the selected subproblems I.
  1: **for** $i = 1$ to N **do**
  2:     $\Delta^i = \frac{f_{main}^{old\_i} - f_{main}^{new\_i}}{f_{main}^{old\_i}}$.
  3:     $\pi^i = \begin{cases} 1, & \text{if } \Delta^i > 0.001 \\ (0.95 + 0.05 \cdot \frac{\Delta^i}{0.001}) \cdot \pi^i, & otherwise \end{cases}$.
  4: **end for**
  5: Set $I = \emptyset$ and select the indices of the m-subproblems whose epsilon vectors are permutation of $(1, 0, \ldots, 0)$. Choose other $\lfloor \frac{N}{5} \rfloor - m - 1$ indices using the 10-tournament selection according to π^i, and add them to I.

---

wasting potentially useful solutions and making the best use of them, the subproblem-to-solution matching procedure is proposed. In this matching procedure, the subproblem that has the minimum constraint violation value is selected for the newly generated solution. The constraint violation value of solution $\mathbf{y}$ regarding the subproblem with ε^l is defined as described in **Algorithm 4**. Since the feasibility rule is adopted to handle constrained

---

Algorithm 4 Subproblem-to-Solution Matching

Require: New generated solution $\mathbf{y}$ and N subproblems with upper bound vectors $\varepsilon^1, \varepsilon^2 \ldots, \varepsilon^{\mathbf{N}}$.

Ensure: The index of the selected subproblem k.

1: **for** $l = 1$ to N **do**

2:    $CV^l = \sum_{j=1, j \neq s}^{m} \max(\frac{y_j - z_j^*}{z_j^{nad} - z_j^*} - \varepsilon_j^l, 0)$.

3:    **if** $CV^l = 0$ **then**

4:       $CV^l = \dfrac{1}{\sum_{j=1, j \neq s}^{m} (\frac{y_j - z_j^*}{z_j^{nad} - z_j^*} - \varepsilon_j^l)}$.

5:    **end if**

6: **end for**

7: $k = argmin_{i=1,\ldots,N}(CV^1, CV^2, \ldots, CV^N)$. // Select the subproblem for which $\mathbf{y}$ is feasible and is nearest to $\mathbf{y}$ in the objective space.

---

subproblems, this procedure is good for convergence. The solution-to-subproblem matching procedure and the subproblem-to-solution matching procedure consider diversity and convergence, respectively.

2.4.6   *Farthest-candidate Approach*

In DMOEA-εC, an external archive population (EP) is maintained in addition to the evolving population. Thus when a new solution is generated, the EP should be updated. And if the number of individuals in EP exceeds S, EP is pruned until its size equals to S.

In many MOEAs, in order to maintain a good spread of obtained non-dominated solutions, several crowded comparison mechanisms have been proposed. In NSGA-II [Deb *et al.* (2002a)], a crowding distance-based comparison mechanism is adopted. Kukkonen and Deb (2006) put forward an improved pruning of non-dominated solutions. This method removes the solution that has the minimum crowding distance value one by one and recalculates the crowding distance value after each removal until the number of the remaining solutions is equal to the population size. In [Wang *et al.* (2007)], a fast and effective method that is based on the crowding distance using the nearest neighborhood of solutions in the Euclidean sense is proposed. However, Chen *et al.* (2015) pointed out that these methods are unable to get a good spread result under some situations and presented a particular case for further explanation. Thus, the farthest-candidate approach [Chen *et al.* (2015)] inspired by the best-candidate sampling algorithm [Mitchell (1991)] in sampling theory is adopted here to prune the EP.

Algorithm 5 Farthest-candidate Approach [Chen *et al.* (2015)]

Require: A population Φ with size F, the size of selected individuals $K(< F)$.
Ensure: The selected solutions P_{accept} with size $K(< F)$.
  1: Initialize $P_{accept} = \emptyset, \mathbf{D} = \mathbf{0}$.
  2: **for** $i = 1$ to m **do**
  3:     $P_{accept} = P_{accept} \cup \underset{\mathbf{x} \in \Phi}{argmin}(f_i(\mathbf{x})) \cup \underset{\mathbf{x} \in \Phi}{argmax}(f_i(\mathbf{x}))$.
  4: **end for**
  5: **for** each $\mathbf{x}$ in $\Phi - P_{accept}$ **do**
  6:     $D(\mathbf{x}) = \underset{\mathbf{x}' \in P_{accept}}{argmin} dist(\mathbf{x}, \mathbf{x}')$.
  7: **end for**
  8: **while** $|P_{accept}| < N$ **do**
  9:     $\mathbf{x_1} = argmax_{\mathbf{x} \in (\Phi - P_{accept})}(D(x))$.
 10:     **for** each $\mathbf{x_2}$ in $\Phi - P_{accept}$ **do**
 11:         $D(\mathbf{x_2}) = \min(D(\mathbf{x_2}), dist(\mathbf{x_1}, \mathbf{x_2}))$.
 12:         $P_{accept} = P_{accept} \cup \mathbf{x_1}$.
 13:     **end for**
 14: **end while**

In the farthest-candidate approach, boundary points (solutions with the minimum and maximum objective values) are selected first. Then the candidate point, which is farthest from the selected points, is chosen iteratively among the unselected points. In this way, a set of evenly distributed non-dominated solutions will be selected from a set of alternative non-dominated solutions. In **Algorithm 5**, P_{accept} stores selected solutions, $\mathbf{D}$ stores the minimum Euclidean distance between $\mathbf{x}$ and unselected points, and $dist(\mathbf{x}, \mathbf{x}')$ is a function that calculates the Euclidean distance between $\mathbf{x}$ and $\mathbf{x}'$. The superiority of the farthest-candidate approach over the crowding distance-based approach used in [Deb *et al.* (2002a)] will be demonstrated in the following.

2.4.7   *Computational Complexity Analysis*

The time complexity analysis of DMOEA-εC is presented in Table 2.1. In summary, the time complexity of DMOEA-εC is $O(m \cdot S^2) \approx O(m \cdot N^2)$. Besides, the time complexity of MOEA/D and MOEA/D-DRA is $O(m \cdot N \cdot T)$ [Zhang and Li (2007); Zhang *et al.* (2009a)]. The time complexity of a MOEA/D-AWA is $O(m \cdot N^2 \cdot (T + nus))$ [Qi *et al.* (2014)]. Compared with a MOEA/D and its variants, DMOEA-εC allocates additional computational resources to the solution-to-subproblem matching procedure, the subproblem-to-solution matching procedure and

Table 2.1   Time Complexity Analysis of DMOEA-εC.

Procedure	Time Complexity
Solution-to-subproblem matching	$O(m \cdot N)$
Extract non-dominated solutions	$O(m \cdot N^2)$
Dynamic resource allocation	$O(N)$
Generate a new solution	$O(m)$
Update neighborhood solutions	$O(m \cdot T)$
Subproblem-to-solution matching	$O(m)$
Update EP using **Algorithm 5**	$O(m \cdot S^2) \approx O(m \cdot N^2)$

the farthest-candidate approach when pruning the EP. The computational resources spent on two matching procedures are negligible. The main time complexity is introduced by the farthest-candidate approach described in **Algorithm 5**.

2.4.8   *Discussions*

2.4.8.1   *Main differences between DMOEA-εC and MOEA/D*

Both DMOEA-εC and MOEA/D introduce the concept of decomposition into MOEAs. Specifically, MOEA/D decomposes an MOP into N scalar subproblems by a scalarizing function. DMOEA-εC selects one of the objectives as the main objective function and converts the other non-main objectives into constraints by giving them upper bound coefficients. Based on N evenly distributed upper bound vectors, an MOP is decomposed into N scalar-constrained subproblems. Similarly, the neighborhood of each subproblem is defined according to the Euclidean distances from the upper bound vector corresponding to the subproblem to other upper bound vectors for DMOEA-εC. N subproblems are optimized using the neighbour information in parallel. However, there are three special mechanisms in DMOEA-εC for improving the performance. Firstly, since DMOEA-εC tends to retain feasible solutions for each subproblem, this will be bad for the optimization of the main objective function. Thus a main objective alternation strategy is proposed. In order to tackle problems induced by the main objective alternation strategy, a solution-to-subproblem matching procedure is proposed to place the nearest solution to each subproblem. Lastly, a subproblem-to-solution matching procedure is used to find a subproblem with the minimum constraint violation value for the newly generated solution.

2.4.8.2   *Main differences between DMOEA-εC and AUGMECON*

Both DMOEA-εC and variants of AUGMECON convert an MOP into a series of scalar-constrained subproblems, but they handle these subproblems in totally different ways. Specifically, variants of AUGMECON optimize them one by one, while DMOEA-εC solves them collaboratively by using the neighbour information. Furthermore, it is worth mentioning that the exact or approximated ideal point and nadir point are needed in advance in various variants of AUGMECON, while there is no such limitation in DMOEA-εC. Actually, the ideal point and the nadir point are updated in each generation.

2.5   Experimental Design

2.5.1   *Multi-objective Continuous Test Instances*

The ZDT test instances, tri-objective DTLZ test instances [Deb *et al.* (2002b)] and UF test suites [Zhang *et al.* (2009b)] (which are part of the CEC2009) MOP test instances, LZ test suites [Li and Zhang (2009)] (with complicated) PF shapes, and bi-objective WFG test suites [Huband *et al.* (2005)] (with complicated PF shapes), are adopted for comparing DMOEA-εC with other six MOEAs.

2.5.2   *Multi-objective 0/1 Knapsack Problems*

This section presents the MOKPs used in the following experiments as benchmarks. Given a set of n items and a set of m knapsacks, the MOKPs can be stated as:

$$\text{maxmize} \quad f_i(\mathbf{x}) = \sum_{j=1}^{n} p_{ij} x_j, i = 1, \ldots, m$$

$$\text{subject to} \quad \sum_{j=1}^{n} w_{ij} x_j \leq c_i, i = 1, \ldots, m \qquad (2.2)$$

$$\mathbf{x} = (x_1, \ldots, x_n) \in \{0, 1\}^n,$$

where $p_{ij} \geq 0$ is the profit of item j in knapsack i, $w_{ij} \geq 0$ is the weight of item j in knapsack i, and c_i is the capacity of knapsack i. $x_i = 1$ means that item i is selected and put into knapsacks.

2.5.3 *Performance Metrics*

Three commonly used performance metrics, i.e., inverted generational distance (*IGD*) [Zhou *et al.* (2005)], hypervolume (*HV*) [Zitzler and Thiele (1999)] and additive ε-indicator ($I_{\epsilon+}$) [Zitzler *et al.* (2003)] are employed to evaluate the performance of all compared algorithms. The MOKPs are NP-hard and can model a variety of applications in resource allocation.

2.5.4 *MOEAs for Comparison*

Six state-of-the-art MOEAs are considered as competitive candidates, including MOEA/D [Zhang and Li (2007)], MOEA/D-DRA [Zhang *et al.* (2009a)], MOEA/D-AWA [Qi *et al.* (2014)], SMEA [Zhang *et al.* (2016)], MOCell [Nebro *et al.* (2009a)], and SMPSO [Nebro *et al.* (2009b)]. MOEA/D, MOEA/D-DRA, and MOEA/D-AWA are all based on decomposition and perform better than a number of popular algorithms. SMEA is a newly proposed competitive multi-objective evolutionary algorithm. It is based on the self-organizing mapping method (SOM) and the neighborhood relationship concept. SMEA has been compared with some advanced multi-objective evolutionary methods and has shown its advantages over competitive approaches. MOCell and SMPSO are cellular-based and particle swarm optimization-based multi-objective solvers, respectively. They both can obtain competitive results on ZDT test suites.

We use the implementation of MOEA/D, MOEA/D-DRA, MOCell and SMPSO provided by the jMetal framework [Durillo and Nebro (2011)].[1] Besides, DMOEA-εC[2] and SMEA are implemented in MATLAB, while MOEA/D-AWA is implemented in C++.[3] All of them are executed on the same computer.

2.5.5 *Parameter Settings*

2.5.5.1 *Public parameter settings*

For a fair comparison, the choice of parameters are the same as the comparison algorithms. Specifically, the population size is set to $N = 100$ for

---

[1]Downloadable from https://github.com/jMetal/jMetal.
[2]The source codes of DMOEA-εC can be downloaded from http://pris.bit.edu.cn/home/people/OtherStaff/xinbin.htm.
[3]The source codes of MOEA/D-AWA and SMEA are obtained from their authors.

the ZDT and bi-objective WFG problems. Due to the differences in algorithmic frameworks, N is set to 351, 306, 324, 300 for the tri-objective DTLZ problem for MOEA/D and its variants, MOCell, DMOEA-εC, and the remaining algorithms, respectively.[4] As for the UF problems, population size is set to $N = 600$ for bi-objective and $N = 1000$ for tri-objective. Since 4 out of 9 LZ test problems are included in the UF test suites, the population size is set to $N = 300$ for the remaining LZ problems. For a fair comparison, an external archive population with the size of $S = \lfloor 1.5N \rfloor$ is added to the comparison algorithms. Besides, the DE operator and Gaussian mutation are used in solving ZDT, DTLZ and WFG test problems. The DE operator and polynomial mutation are adopted in solving UF and LZ test problems. Moreover, control parameters for these reproduction operators are the same as those claimed in comparative MOEAs.[5] All compared algorithms stop when the number of function evaluations reaches the maximum number. For a fair comparison, in accordance with the parameter settings in comparison algorithms, the maximum number of function evaluations is set to 50,000 for the ZDT problems, 75,000 for the tri-objective DTLZ problems, 300,000 for the UF problems, 150,000 for the remaining LZ problems, and 45,000 for the bi-objective WFG problems. Finally, each algorithm is executed 30 times independently on each instance.

2.5.5.2 *Parameter settings in MOEA/D, MOEA/D-DRA and MOEA/D-AWA*

Parameter settings adopted here are the same as those claimed in [Qi *et al.* (2014); Zhang and Li (2007); Zhang *et al.* (2009a)].

Neighborhood size: $T = \lfloor 0.1N \rfloor$;
Probability of selecting mate solutions from neighborhood: $\delta = 0.9$; and
Maximal number of replacement: $n_r = \lfloor 0.01N \rfloor$.

---

[4]Since the algorithmic frameworks of the proposed DMOEA-εC and comparison algorithms are all different, the calculations of the population size N are conducted differently. The population size of MOEA/D is determined by the number of weight vectors $N = C_{H+m-1}^{m-1}$ (m is the number of objectives and H is a controlled parameter). In order to have a comparable population size, for the tri-objective problems we set $H = 25$, thus $N = 351$. The population size of MOCell is determined by the number of cellular grids. For the tri-objective problems we set $N = 17 \times 18 = 306$. The population size of DMOEA-εC is determined by $N = q^{m-1}$ (q is a controlled parameter). For the tri-objective problems we set $q = 18$, thus $N = 324$. For the remaining algorithms, we set $N = 300$.

[5]CR is changed to 0.9 only for ZDT problems.

In addition to the above-mentioned common parameters, the iteration interval of utilizing the dynamic resource allocation strategy is set as $DRA\_interval = 50$ for MOEA/D-DRA and MOEA/D-AWA. For MOEA/D-AWA, the maximal number of subproblems adjusted is set as $nus = \lfloor 0.05N \rfloor$. The parameter $rate\_evol$ is set to 0.8.

2.5.5.3   *Parameter settings in SMEA*

Parameter settings in SMEA adopted here are the same as those claimed in [Zhang *et al.* (2016)].

SOM structures: 1-dimensional structure 1×100 for bi-objective MOPs;
Initial learning rate: $\tau_0 = 0.9$;
Neighborhood size: $T = 5$;
Probability of selecting mate solutions from the neighborhood $\delta = 0.7$; and
Maximal number of replacement: $n_r = 1$.

2.5.5.4   *Parameter settings in MOCell*

Parameter settings in MOCell adopted here are the same as those claimed in [Nebro *et al.* (2009a)].

Neighborhood: 1-hop neighbours (8 surrounding solutions);
Selection of parents: binary tournament + binary tournament; and
Feedback: 20 individuals.

2.5.5.5   *Parameter settings in SMPSO*

Parameter settings in SMPSO adopted here are the same as those claimed in [Nebro *et al.* (2009b)].

The inertia weight: $w = 0.1$; and
The range of C_1 and C_2: $[1.5, 2.5]$.

2.5.5.6   *Parameter settings in DMOEA-εC*

When compared with each algorithm, parameter settings in DMOEA-εC are set the same as each competitor. Besides, the setting of IN_m varies with different test problems. IN_m is set to $\lfloor 50\% \cdot (\text{number of iterations}) \rfloor$ for ZDT problems, $\lfloor 20\% \cdot (\text{number of iterations}) \rfloor$ for UF, LZ and WFG problems, and $\lfloor 10\% \cdot (\text{number of iterations}) \rfloor$ for DTLZ problems based on the parameter sensitivity analysis.

2.6 Numerical Results on Continuous and Discrete Test Instances

This section is devoted to the experimental design for investigating the performance of DMOEA-εC on continuous test instances and discrete MOKPs.

2.6.1 *Comparisons on Multi-objective Continuous Test Instances*

This part of the experiment is designed to study the effectiveness of DMOEA-εC on continuous MOPs. At first, the classical ZDT and DTLZ test suites are investigated. The performance of DMOEA-εC on more complicated test instances will be studied later.

In calculating the performance metrics, 100 non-dominated solutions selected from the combination of the evolving population and EP using the farthest-candidate approach (**Algorithm 5**) is used in the case of ZDT and WFG problems, and 300 in the case of DTLZ problems. Similarly, for UF and LZ test problems, 100 and 150 non-dominated solutions are selected and used for the performance metrics calculation for two-objective and tri-objective problems, respectively.

With the purpose of calculating the *IGD* metric value, P^* is chosen to be a set of 500 uniformly distributed points along the true PF for ZDT problems, and 1024 points for DTLZ instances. As to two-objective UF and LZ problems, a set of 1000 uniformly distributed points along the true PF are chosen as P^*, except that 21 uniformly distributed points are chosen as P^* for UF5. And for tri-objective UF test problems, P^* is chosen to be a set of 10,000 uniformly distributed points along the true PF. The P^* that was used for computing *IGD* metrics for WFG problems is the same as in [Zhang *et al.* (2016)].

Besides, in order to compute the *HV* metric value, the reference point is set as 1.1 times the true nadir point. Specifically, reference points of different test instances are illustrated in Table 2.2.

Table 2.2   Reference Points of Test Instances.

Instance	Reference Point
ZDT1-ZDT4, ZDT6, UF1-UF7, LZ1, LZ3, LZ4, LZ7, LZ9	(1.1, 1.1)
DLTZ1-DTLZ4, UF8-UF10	(1.1, 1.1, 1.1)
DTLZ6	(1.1, 1.1, 6.6)
WFG1-WFG9	(2.2, 4.4)

The means and standard deviations of *IGD*, *HV* and $I_{\epsilon+}$ metric values over 30 independent runs of each algorithm on 34 test instances are shown in Tables 2.3 to 2.5, respectively. The mean *HV* (*IGD*/ $I_{\epsilon+}$) values for each instance are sorted in descending (ascending) order, and the numbers in the square brackets are their ranks. The Wilcoxon's rank sum test at a 5% significance level is conducted to test the significance of differences between the mean metric values yielded by DMOEA-εC and comparison algorithms. The numbers in parentheses are the standard deviations. †, § and ≈ indicate that the performance of DMOEA-εC is better than, worse than, or similar to that of the comparison algorithm according to the Wilcoxon's rank sum test, respectively. The bold data in the table are the best mean metric values for each instance. Besides, Table 2.6 summarizes the overall performance, including the mean ranks and statistical results obtained via the Wilcoxon's rank sum test, of 7 algorithms on 34 instances, in terms of 3 metric values.

2.6.1.1 *Experimental results on ZDT and DTLZ test suites*

The mathematical descriptions of these test problems and true PFs can be found in [Zitzler *et al.* (2000)] and [Deb *et al.* (2002b)].

As can be seen in Tables 2.3 to 2.5, in terms of *IGD* metric values, SMPSO shows a significant advantage over DMOEA-εC on ZDT test problems, and the performance of DMOEA-εC is no worse than that of comparison algorithms on all DTLZ problems except the DTLZ1 and DTLZ2, on which MOEA/D and MOEA/D-AWA show better performance, respectively. For the *HV*, DMOEA-εC shows significant superiority over others on both ZDT and DTLZ problems except ZDT3, ZDT4 and DTLZ2. As to $I_{\epsilon+}$, SMPSO outperforms DMOEA-εC significantly on all ZDT problems, and DMOEA-εC shows a clear advantage over others on all DTLZ problems. Besides, both DMOEA-εC and MOEA/D-AWA perform better than any other algorithms on the majority of DTLZ problems in terms of all 3 metrics. The difference lies in that DMOEA-εC tends to obtain solutions with good convergence, whereas MOEA/D-AWA does well in maintaining good uniformity. In summary, the performance of DMOEA-εC and variants of MOEA/D on ZDT test instances is not as promising as SMPSO. However, the superiority of the proposed DMOEA-εC over comparison algorithms is highlighted on DTLZ problems.

Table 2.3   Statistical Results of 7 Algorithms over 30 Independent Runs on the 34 Instances in Terms of *IGD* Metrics.

Instance	MOEA/D	MOEA/D-DRA	MOEA/D-AWA	SMEA	MOCell	SMPSO	DMOEA-εC
ZDT1	3.800E-03†[4](0.000E-07)	4.903E-03†[8](8.829E-04)	4.430E-03†[5](7.940E-05)	2.218E-02†[7](7.060E-03)	3.700E-03**[2](2.631E-05)	**3.653E-03§[1](5.07E-05)**	3.763E-03[3](6.690E-05)
ZDT2	4.000E-03†[4](1.830E-05)	5.760E-03†[6](1.388E-03)	4.463E-03†[5](6.690E-05)	3.799E-02†[7](1.393E-02)	3.800E-03**[2](3.710E-05)	**3.787E-03§[1](3.460E-05)**	3.800E-03[3](7.120E-05)
ZDT3	1.787E-02†[6](1.255E-02)	1.101E-02†[5](5.683E-04)	5.867E-03†[3](1.422E-04)	3.587E-02†[7](2.713E-02)	7.293E-03†[4](9.122E-03)	**4.313E-03§[1](3.457E-05)**	5.141E-03[2](1.696E-04)
ZDT4	3.890E-03‡[2](7.588E-05)	3.722E-02†[6](5.545E-02)	4.173E-03†[7](2.518E-04)	3.110E-00†[7](1.745E-00)	3.953E-03**[3](2.932E-04)	**3.707E-03§[1](2.537E-05)**	4.060E-03[4](2.799E-05)
ZDT6	3.330E-03†[6](0.000E-07)	3.100E-03†[5](8.820E-05)	3.056E-02†[3](8.764E-04)	3.0003E-03**[4](1.825E-05)	3.000E-03**[2](1.323E-05)	**3.000E-03§[1](1.322E-05)**	3.000E-03[3](8.050E-05)
DTLZ1	5.312E-02†[6](1.111E-03)	2.376E-02‡[1](2.596E-04)	2.998E-02‡[1](4.145E-04)	8.434E-01†[7](1.004E-00)	3.819E-02†[5](5.539E-03)	3.396E-02[4](8.261E-04)	2.476E-02[2](9.966E-04)
DTLZ2	6.684E-02†[6](2.845E-04)	3.075E-02*[2](3.893E-04)	3.131E-02†[2](1.494E-03)	7.012E-02†[7](7.213E-04)	3.935E-02†[4](1.218E-03)	4.014E-02[5](1.299E-03)	3.160E-02[2](8.819E-04)
DTLZ3	6.697E-02†[5](5.616E-04)	5.655E-02†[4](1.089E-01)	2.997E-02*‡[1](2.578E-04)	3.467E-00†[7](5.957E-00)	8.563E-02†[6](2.776E-02)	5.336E-02[3](5.332E-02)	**3.068E-02§[1](7.981E-04)**
DTLZ4	6.703E-02†[6](8.167E-04)	4.963E-02†[4](6.899E-02)	3.334E-02*‡[1](6.906E-04)	2.154E-01†[7](8.791E-02)	3.670E-02†[3](8.424E-04)	6.267E-02[5](3.097E-02)	3.020E-02[2](1.021E-03)
DTLZ6	3.875E-01†[7](1.834E-01)	1.578E-02†[5](1.987E-01)	5.700E-03†[4](1.907E-04)	7.852E-02†[7](5.004E-02)	2.044E-01†[6](2.369E-01)	4.541E-02[3](2.876E-03)	3.360E-02[2](5.777E-04)
UF1(L22)	9.397E-03†[5](1.623E-03)	4.423E-03*[2](9.610E-05)	8.660E-03†[3](1.322E-03)	5.123E-03†[3](9.497E-04)	1.059E-01†[7](2.650E-02)	6.175E-02[6](9.723E-03)	**4.407E-03§[1](9.870E-05)**
UF2(L25)	1.436E-02†[5](3.333E-03)	6.227E-03*‡[1](1.205E-03)	1.046E-02†[3](6.849E-03)	1.151E-02†[4](2.263E-03)	4.937E-02†[7](2.582E-02)	2.378E-02[6](1.930E-03)	6.503E-03[2](5.229E-04)
UF3(L28)	2.107E-02†[5](1.990E-02)	7.951E-03†[3](5.675E-03)	6.469E-02†[5](3.061E-03)	1.458E-02†[4](9.682E-03)	3.032E-01†[7](2.079E-02)	1.118E-01[6](3.447E-02)	**7.209E-03§[1](1.355E-02)**
UF4	6.907E-02†[6](7.412E-03)	6.078E-02*[4](4.158E-03)	1.224E-01*‡[1](2.519E-01)	9.769E-02†[7](8.374E-03)	**4.503E-02*‡[1](1.756E-03)**	5.142E-02*[2](2.829E-03)	6.017E-02[3](5.017E-03)
UF5	3.355E-01†[3](1.409E-01)	3.190E-01†[3](1.228E-01)	5.608E-01†[4](2.930E-01)	3.941E-01†[6](8.814E-02)	3.452E-01**[4](9.993E-02)	1.872E-00[7](5.442E-01)	3.845E-01[5](1.251E-01)
UF6	3.096E-01†[5](2.215E-01)	1.765E-01†[3](1.124E-01)	6.430E-03†[4](3.831E-04)	1.024E-01*‡[1](5.955E-02)	4.309E-01†[1](1.990E-01)	4.484E-01[7](9.853E-02)	2.662E-02[4](2.012E-01)
UF7	7.633E-03†[5](3.333E-03)	4.435E-03*[2](1.537E-04)	1.257E-01†[4](3.783E-02)	6.413E-03†[3](6.066E-04)	2.954E-01†[7](1.627E-01)	2.253E-02[6](2.439E-03)	**4.079E-03§[1](1.269E-03)**
UF8(L26)	7.988E-02†[3](1.389E-02)	6.350E-02†[2](1.144E-02)	1.677E-01†[4](2.248E-02)	2.004E-01†[6](8.619E-02)	2.079E-01†[7](5.437E-02)	1.911E-01[5](2.907E-02)	**5.284E-02§[1](1.058E-02)**
UF9	1.353E-01†[3](4.758E-02)	1.034E-01†[2](5.370E-02)	4.833E-01†[8](4.563E-02)	1.694E-01†[5](8.748E-02)	2.067E-01†[7](6.252E-02)	1.911E-01[6](2.077E-02)	**4.292E-02§[1](1.660E-02)**
UF10	4.490E-01†[5](7.552E-02)	4.249E-01†[4](7.563E-02)	1.466E-03†[4](8.997E-05)	1.039E-00†[7](2.187E-01)	4.151E-01†[3](1.108E-01)	2.756E-01[2](2.739E-02)	**2.240E-02§[1](7.361E-02)**
L21	**1.290E-03*‡[1](3.724E-05)**	1.664E-02†[5](9.812E-05)	3.806E-03*[2](4.472E-05)	3.613E-03†[7](2.342E-05)	2.367E-03†[6](1.454E-04)	1.431E-03[3](3.391E-05)	1.395E-03[2](1.230E-03)
L23	6.122E-03†[4](3.810E-03)	3.693E-03†[3](6.772E-04)	3.806E-03*[2](4.472E-05)	7.129E-03†[5](2.560E-03)	6.570E-02†[7](2.978E-02)	3.627E-02[6](5.754E-03)	**2.372E-03§[1](2.794E-03)**
L24	7.265E-03†[5](3.758E-03)	3.075E-02†[3](4.275E-04)	2.733E-03†[2](1.211E-04)	3.979E-03†[4](1.201E-04)	4.570E-02†[7](4.098E-03)	3.740E-02[6](3.865E-03)	**1.977E-03§[1](3.780E-03)**
L27	3.056E-02†[5](4.478E-02)	2.599E-02†[3](4.286E-04)	2.040E-03†[2](2.608E-04)	4.065E-03†[4](1.851E-04)	4.070E-01†[7](1.2453E-01)	1.083E-01[6](5.835E-02)	**1.479E-03§[1](2.691E-03)**
L29	9.770E-03†[3](4.729E-03)	4.220E-03†[2](6.390E-04)	7.264E-02†[5](5.764E-03)	1.504E-02†[4](3.084E-03)	1.475E-01†[7](6.661E-02)	7.330E-02[6](1.615E-02)	**3.192E-02§[1](4.437E-02)**
WFG1	1.202E-00†[4](2.487E-02)	1.206E-00†[5](2.728E-02)	1.181E-00†[3](3.451E-02)	1.555E-00†[7](1.003E-01)	9.814E-01*‡[1](1.804E-01)	1.237E-00[6](5.391E-03)	1.063E-00[2](1.575E-02)
WFG2	7.922E-02†[5](1.301E-02)	7.440E-02†[3](1.660E-02)	1.336E-01†[7](6.518E-02)	**1.492E-02*‡[1](9.970E-04)**	7.815E-02[4](2.234E-02)	8.653E-02[6](1.651E-02)	1.505E-02[2](8.403E-04)
WFG3	4.259E-02†[5](7.320E-03)	3.449E-02†[3](5.147E-03)	6.601E-02†[6](2.324E-02)	**1.182E-02*‡[1](1.177E-04)**	3.726E-02[4](1.399E-02)	6.742E-02[7](1.566E-02)	1.391E-01[2](1.199E-03)
WFG4	1.063E-01†[7](8.903E-03)	9.827E-02†[6](9.393E-03)	3.355E-02†[2](9.908E-03)	8.401E-02[5](5.844E-03)	**1.862E-02*‡[1](1.942E-03)**	7.037E-02[4](2.725E-03)	4.008E-02[3](7.401E-03)
WFG5	6.749E-02†[5](3.991E-04)	6.796E-02†[6](8.381E-04)	6.812E-02†[7](2.810E-04)	6.662E-02[3](1.053E-04)	6.694E-02[7](1.612E-05)	6.637E-02**[2](5.960E-03)	**6.633E-02§[1](7.970E-04)**
WFG6	8.576E-02†[7](1.365E-02)	8.278E-02†[6](1.747E-02)	5.481E-02†[5](1.633E-02)	**1.356E-02*‡[1](3.048E-04)**	5.209E-02[4](9.408E-03)	2.614E-02**[3](1.916E-03)	2.576E-02[2](2.087E-04)
WFG7	2.417E-02†[6](1.276E-03)	2.310E-02†[5](1.297E-03)	1.935E-02†[3](1.705E-03)	1.144E-02[2](2.596E-04)	1.309E-02[4](2.303E-04)	3.906E-02[7](7.782E-03)	**1.052E-02§[1](7.213E-04)**
WFG8	1.268E-01†[5](9.854E-03)	1.230E-01†[4](1.360E-02)	1.431E-01†[6](6.412E-02)	**3.323E-02*‡[1](8.009E-03)**	8.057E-02[3](7.866E-03)	1.442E-01[7](1.630E-02)	3.604E-02[2](4.116E-03)
WFG9	8.308E-02†[7](1.802E-02)	7.564E-02†[6](2.411E-02)	5.477E-02†[5](2.333E-02)	2.900E-02[2](4.801E-02)	3.723E-02[4](2.558E-02)	**1.404E-02*‡[1](3.439E-04)**	3.054E-02[3](1.664E-03)
†/§/≈	31/2/1	26/4/4	26/4/4	27/4/3	26/2/6	25/6/3	25/6/3

Table 2.4 Statistical Results of 7 Algorithms over 30 Independent Runs on the 34 Instances in Terms of *HV* Metrics.

Instance	MOEA/D	MOEA/D-DRA	MOEA/D-AWA	SMEA	MOCell	SMPSO	DMOEA-εC
ZDT1	8.686E-01[4](1.541E-02)	8.679E-01[5](8.407E-03)	8.710E-01[3](1.607E-02)	8.380E-01[6](1.385E-02)	8.710E-01[2](6.023E-03)	7.922E-01[7](6.584E-03)	**8.727E-01[1](1.541E-02)**
ZDT2	5.421E-01≈[2](1.091E-02)	5.363E-01[6](7.542E-03)	5.421E-01≈[3](1.751E-02)	4.783E-01[7](2.546E-02)	5.393E-01[4](8.484E-03)	5.389E-01[5](9.047E-03)	**5.434E-01[1](1.405E-02)**
ZDT3	1.003E-00≈[4](1.124E-02)	9.988E-01[6](6.994E-03)	1.006E-00[3](1.111E-02)	9.760E-01[7](2.672E-02)	**1.008E-00[1](6.811E-03)**	1.006E-00[2](7.988E-03)	1.003E-00[5](1.304E-02)
ZDT4	**8.896E-01[1](1.392E-02)**	8.087E-01[6](8.906E-03)	8.660E-01≈[5](1.121E-02)	1.575E-01[7](1.643E-01)	8.705E-01[3](6.904E-03)	8.707E-01[2](8.279E-03)	8.668E-01[4](4.691E-02)
ZDT6	5.042E-01[5](1.635E-02)	5.045E-01[4](7.675E-03)	4.974E-01≈[7](1.919E-02)	5.055E-01[2](9.793E-03)	5.037E-01[6](8.735E-03)	5.054E-01[3](7.660E-03)	**5.086E-01[1](1.469E-02)**
DTLZ1	7.062E-01[7](1.742E-02)	1.306E-00≈[2](2.233E-03)	1.122E-00[3](5.211E-02)	9.318E-01[6](3.779E-01)	1.111E-00[5](1.341E-02)	1.120E-00[4](6.384E-03)	**1.307E-00[1](4.458E-03)**
DTLZ2	7.124E-01[7](1.232E-02)	**7.717E-01[1](9.340E-03)**	7.773E-01[2](1.373E-02)	7.591E-01[3](7.929E-03)	7.406E-01[5](1.041E-02)	7.351E-01[6](1.226E-02)	7.435E-01[4](1.528E-02)
DTLZ3	7.006E-01[4](1.572E-02)	7.295E-01[1](1.434E-01)	7.596E-01[2](1.906E-02)	6.628E-01[6](1.511E-03)	6.745E-01[5](1.185E-01)	6.515E-01[7](7.736E-02)	**7.696E-01[1](1.372E-02)**
DTLZ4	7.109E-01[6](1.528E-02)	7.618E-01[3](3.022E-02)	7.701E-01[2](4.222E-02)	6.830E-01[7](4.605E-02)	7.295E-01[5](1.155E-01)	7.372E-01[4](1.627E-02)	**7.840E-01[1](1.656E-02)**
DTLZ6	2.691E-00[7](2.924E-02)	2.412E-00[7](2.683E-01)	2.647E-00[2](1.561E-02)	2.609E-00[3](1.131E-01)	2.348E-00[6](3.287E-01)	2.582E-00[4](5.797E-02)	**3.118E-00[1](3.344E-02)**
UF1(LZ2)	8.599E-01[5](9.680E-03)	8.670E-01[4](1.759E-01)	8.675E-01[3](6.371E-03)	8.697E-01[2](5.929E-03)	7.104E-01[7](3.601E-02)	7.625E-01[6](2.377E-02)	**8.741E-01[1](1.434E-02)**
UF2(LZ5)	8.554E-01[5](9.815E-03)	8.675E-01≈[3](7.208E-02)	8.614E-01[3](7.108E-03)	8.591E-01[4](6.946E-03)	8.257E-01[7](2.319E-02)	8.397E-01[6](8.459E-02)	**8.700E-01[1](1.382E-02)**
UF3(LZ8)	8.428E-01[5](2.811E-02)	8.502E-01[4](4.456E-02)	8.546E-01≈[3](1.767E-02)	**1.039E-00[1](8.405E-03)**	4.617E-01[7](3.578E-02)	7.072E-01[6](4.944E-02)	8.687E-01[2](1.633E-02)
UF4	4.234E-01[6](1.298E-02)	4.371E-01[4](1.611E-02)	4.322E-01[5](9.799E-03)	3.717E-01[7](1.666E-02)	**4.698E-01[1](5.862E-03)**	4.536E-01≈[2](9.896E-03)	4.436E-01[3](1.965E-02)
UF5	2.206E-01[3](9.592E-02)	2.749E-01[3](8.351E-02)	**4.961E-01[1](3.461E-02)**	1.288E-01[5](8.754E-02)	2.949E-01[4](7.512E-02)	3.999E-01[7](6.088E-03)	1.256E-01[6](4.731E-02)
UF6	3.376E-01[3](1.055E-01)	3.170E-00≈[6](1.194E-01)	**5.055E-01[1](6.352E-03)**	4.263E-01[5](6.889E-02)	3.269E-01[4](1.585E-01)	1.068E-01[7](7.217E-02)	3.170E-01[5](1.061E-01)
UF7	6.939E-01[5](9.662E-03)	**7.056E-01≈[1](1.520E-02)**	7.024E-01[3](8.854E-03)	6.983E-01[4](7.392E-03)	4.227E-01[7](1.279E-01)	6.693E-01[6](8.930E-03)	7.054E-01[2](1.701E-02)
UF8(LZ6)	6.558E-01[3](2.293E-02)	**7.250E-01[1](1.602E-02)**	5.618E-01[5](5.928E-02)	6.050E-01[4](5.928E-02)	4.649E-01[7](1.407E-01)	4.323E-01[7](5.121E-02)	7.025E-01[2](2.423E-02)
UF9	8.901E-01[3](7.054E-02)	9.149E-01[4](4.842E-02)	8.252E-01[5](3.913E-02)	8.582E-01[4](7.610E-02)	7.171E-01[6](1.135E-01)	5.109E-01[7](1.118E-01)	**1.019E-00[1](5.058E-02)**
UF10	1.446E-01[3](4.563E-02)	1.956E-01[3](3.904E-02)	1.377E-01[6](2.136E-02)	3.284E-01[7](5.632E-03)	1.618E-01[4](5.557E-02)	3.171E-01[2](4.974E-02)	**6.088E-01[1](9.056E-02)**
LZ1	8.762E-01≈[2](1.136E-03)	8.732E-01[6](8.066E-03)	8.747E-01[4](7.732E-02)	8.709E-01[3](9.450E-03)	8.746E-01[3](1.011E-02)	8.749E-01[3](1.011E-02)	**8.797E-01[1](4.100E-03)**
LZ3	8.664E-01[4](9.613E-03)	8.701E-01[3](1.332E-02)	8.738E-01[2](1.197E-02)	8.661E-01[5](8.253E-03)	7.802E-01[6](2.010E-02)	8.250E-01[6](1.354E-02)	**8.760E-01[1](8.710E-03)**
LZ4	8.647E-01[5](8.462E-03)	8.697E-01[3](5.884E-03)	8.763E-01≈[2](2.063E-03)	8.712E-01[3](7.096E-03)	8.105E-01[7](8.746E-03)	8.193E-01[6](9.111E-03)	**8.794E-01[1](4.100E-03)**
LZ7	7.900E-01[5](8.856E-02)	8.666E-01[4](8.494E-03)	8.761E-01[2](1.045E-02)	8.698E-01[3](8.372E-03)	4.493E-01[7](8.731E-03)	6.456E-01[6](1.022E-01)	**8.806E-01[1](1.350E-02)**
LZ9	5.213E-01[3](1.818E-02)	**5.340E-01[1](7.182E-03)**	4.207E-01[5](5.506E-03)	5.118E-01[4](1.219E-02)	3.573E-01[7](4.096E-02)	4.116E-01[6](1.963E-02)	5.322E-01[2](2.514E-02)
WFG1	1.787E-00[5](6.072E-02)	1.803E-00[4](8.322E-02)	0.0000E-07[7](0.0000E-07)	6.271E-00[2](2.483E-01)	3.672E-00[3](3.882E-01)	1.620E-00[6](6.651E-02)	**8.616E-00[1](1.512E-02)**
WFG2	5.843E-00[4](7.759E-02)	5.905E-00[3](9.227E-02)	5.531E-00[7](2.657E-01)	6.025E-00≈[2](1.853E-02)	5.799E-00[1](1.063E-01)	5.652E-00[6](8.879E-02)	**6.076E-00[1](1.311E-02)**
WFG3	5.430E-00[5](1.741E-02)	5.464E-00≈[3](5.647E-02)	5.253E-00[7](1.693E-01)	**5.612E-01[1](1.874E-02)**	5.455E-00[4](1.207E-01)	5.298E-00[6](1.178E-01)	5.472E-00[2](1.541E-02)
WFG4	2.858E-00[7](7.019E-02)	2.907E-00[6](7.325E-02)	3.242E-00[8](8.660E-02)	2.984E-00[5](2.042E-02)	**3.415E-00[1](1.564E-01)**	3.007E-00[4](5.869E-02)	3.105E-00[3](1.645E-02)
WFG5	2.981E-00[4](5.306E-02)	2.977E-00[5](7.896E-02)	2.923E-00[7](6.573E-02)	3.003E-00[3](1.473E-02)	**3.006E-00[1](6.018E-02)**	2.971E-00[6](5.716E-02)	3.025E-00[1](1.410E-02)
WFG6	2.929E-00[7](1.117E-01)	2.947E-00[6](1.220E-01)	3.033E-00[5](1.349E-01)	**3.373E-00[1](6.159E-02)**	3.124E-00[4](8.839E-02)	3.251E-00[3](2.279E-02)	3.303E-00[2](2.479E-02)
WFG7	3.285E-00[6](6.356E-02)	3.293E-00[4](7.899E-02)	3.285E-00[5](5.917E-01)	3.334E-00[3](1.670E-02)	3.351E-00[2](7.057E-02)	3.188E-00[7](8.228E-02)	**3.351E-00[1](1.622E-02)**
WFG8	2.692E-00[5](7.609E-02)	2.719E-00[4](7.545E-02)	2.513E-00[7](2.350E-01)	3.228E-00[2](1.672E-02)	2.919E-00[3](7.276E-02)	2.584E-00[6](1.027E-01)	**3.340E-00[1](1.605E-02)**
WFG9	2.911E-00[7](1.347E-01)	2.932E-00[6](1.460E-01)	3.000E-00[5](1.259E-01)	3.230E-00≈[2](2.468E-01)	**3.128E-00[1](1.389E-01)**	**3.261E-00[1](4.866E-02)**	3.138E-00[3](2.067E-02)
↑/↓/≈	28/3/3	25/4/5	24/4/5	26/5/3	25/7/2	30/3/1	

Table 2.5   Statistical Results of 7 Algorithms over 30 Independent Runs on the 34 Instances in Terms of $I_{\epsilon+}$ Metrics.

Reference point

Instance	MOEA/D	MOEA/D-DRA	MOEA/D-AWA	SMEA	MOCell	SMPSO	DMOEA-εC
ZDT1	7.087E-03[4](1.456E-04)	1.061E-02[6](2.261E-03)	8.453E-03[5](3.674E-04)	2.452E-02[1](6.146E-03)	5.457E-03[2](5.361E-04)	**5.270E-03[1](1.643E-04)**	5.700E-03[3](8.120E-04)
ZDT2	6.057E-03[4](5.683E-05)	1.047E-02[6](3.972E-03)	9.193E-03[5](4.770E-04)	7.655E-02[7](3.144E-02)	5.683E-03[2](1.890E-04)	**5.253E-03[1](1.224E-04)**	5.900E-03[3](7.784E-04)
ZDT3	9.549E-02[6](1.349E-01)	1.894E-02[2](3.546E-03)	1.093E-02[2](1.344E-03)	1.180E-01[7](6.644E-02)	3.601E-02[5](9.464E-02)	**5.090E-03[1](2.695E-04)**	2.101E-02[4](1.221E-02)
ZDT4	7.897E-03[5](1.261E-03)	7.667E-02[4](9.489E-02)	8.273E-03[6](3.694E-04)	3.397E-00[7](1.774E-00)	6.153E-03[2](6.426E-04)	**5.467E-03[1](1.988E-04)**	6.600E-03[3](1.994E-03)
ZDT6	4.900E-03[5](8.822E-05)	5.167E-03[6](3.315E-04)	8.683E-03[7](5.025E-04)	4.640E-03[3](2.027E-04)	**4.477E-03[1](1.501E-04)**	4.610E-03[2](2.123E-04)	4.747E-03[4](2.740E-03)
DTLZ1	1.181E-01[6](7.448E-02)	4.287E-02[2](2.819E-03)	5.317E-02[3](5.187E-03)	9.866E-01[7](1.162E-04)	9.002E-02[5](1.904E-02)	7.102E-02[4](4.549E-03)	**4.198E-02[1](3.655E-03)**
DTLZ2	1.381E-01[7](3.326E-03)	4.932E-02[4](1.382E-03)	3.871E-02[2](3.434E-03)	4.684E-02[3](3.509E-03)	8.159E-02[5](1.215E-02)	8.083E-02[5](1.102E-02)	**2.584E-02[1](6.647E-03)**
DTLZ3	1.361E-01[4](8.566E-03)	5.635E-01[6](5.045E-02)	4.253E-02[2](4.258E-03)	4.369E-00[7](3.322E-00)	2.095E-01[5](2.598E-01)	1.129E-01[3](1.282E-01)	**3.997E-02[1](3.283E-03)**
DTLZ4	1.198E-00[7](7.085E-01)	5.076E-01[5](7.204E-01)	3.974E-02[2](5.793E-03)	1.975E-01[4](9.000E-02)	1.011E-01[5](1.699E-01)	8.883E-02[3](2.334E-02)	**3.88E-02[1](2.764E-03)**
DTLZ6	2.014E-02[3](6.884E-03)	5.123E-03[2](1.209E-03)	1.163E-02[3](1.165E-03)	1.033E-01[4](2.193E-01)	2.169E-01[7](5.096E-02)	1.269E-01[6](2.582E-02)	**5.395E-02[1](6.029E-03)**
UF1(LZ2)	6.313E-02[5](1.510E-02)	2.634E-02[2](8.849E-03)	2.756E-02[3](3.562E-03)	1.190E-02[4](6.871E-03)	1.464E-01[7](6.597E-02)	6.684E-02[6](6.506E-03)	**3.680E-03[1](3.326E-04)**
UF2(LZ5)	7.766E-02[5](7.086E-02)	**1.737E-02[1](2.098E-02)**	2.706E-02[3](3.182E-02)	4.924E-02[4](1.432E-02)	5.367E-01[7](8.431E-02)	1.255E-01[6](4.086E-02)	**2.613E-02[1](4.614E-03)**
UF3(LZ8)	8.050E-02[6](5.829E-03)	7.650E-02[4](7.211E-03)	7.685E-02[3](6.577E-03)	6.400E-02[4](2.425E-02)	**4.951E-02[1](7.064E-03)**	6.254E-02[2](7.117E-03)	4.858E-02[3](7.014E-02)
UF4	5.107E-01[3](1.079E-01)	4.754E-01[2](1.538E-01)	**2.535E-01[1](2.657E-02)**	1.106E-01[7](7.107E-03)	6.310E-01[5](1.667E-01)	1.335E-00[7](3.920E-01)	7.123E-02[3](6.624E-03)
UF5	5.275E-01[3](2.421E-01)	5.456E-01[5](3.076E-01)	3.005E-01[5](1.534E-04)	5.795E-01[4](1.504E-01)	6.239E-01[7](2.776E-01)	5.685E-01[6](1.096E-01)	6.389E-01[6](1.932E-01)
UF6	4.689E-02[5](3.799E-01)	**1.587E-02[1](2.044E-02)**	2.506E-02[3](3.988E-03)	**2.919E-01[1](1.441E-01)**	5.893E-01[7](2.294E-01)	7.425E-02[6](1.749E-02)	5.346E-01[4](2.961E-01)
UF7	2.423E-01[3](2.493E-02)	**1.526E-02[2](1.685E-02)**	3.465E-01[5](1.909E-01)	4.558E-02[4](5.989E-03)	6.434E-01[7](1.967E-01)	6.047E-01[6](1.795E-01)	2.304E-02[2](2.342E-02)
UF8(LZ6)	3.628E-01[3](1.126E-01)	3.793E-01[4](1.291E-01)	4.090E-01[6](7.238E-02)	2.489E-01[4](1.339E-01)	3.962E-01[7](1.563E-01)	5.152E-01[7](7.479E-02)	1.956E-01[3](3.391E-02)
UF9	8.337E-01[5](8.264E-02)	7.768E-01[3](7.043E-02)	8.167E-01[4](6.870E-02)	3.237E-01[1](7.674E-01)	8.737E-01[6](7.471E-02)	7.322E-01[2](4.031E-02)	**1.008E-01[1](1.408E-02)**
UF10	2.783E-03[2](5.533E-04)	5.815E-03[3](1.475E-03)	3.186E-03[4](5.178E-04)	1.134E-00[7](1.288E-01)	5.804E-02[3](1.333E-03)	**2.596E-03[1](1.622E-04)**	**3.241E-01[1](1.387E-01)**
LZ1	4.121E-02[3](2.267E-02)	1.622E-02[3](6.724E-03)	**4.581E-03[1](1.789E-04)**	5.237E-02[3](5.099E-04)	1.800E-01[7](6.533E-02)	9.367E-02[1](1.364E-02)	2.883E-03[3](3.400E-03)
LZ3	5.198E-02[5](2.660E-02)	1.257E-02[4](3.712E-03)	6.200E-03[2](5.215E-04)	2.774E-02[4](1.501E-02)	1.721E-01[7](1.806E-02)	1.027E-01[6](1.312E-02)	4.949E-02[2](1.264E-02)
LZ4	1.542E-01[5](1.388E-01)	1.866E-02[4](6.258E-03)	5.180E-03[3](8.927E-04)	7.834E-03[3](1.008E-03)	6.122E-01[7](1.402E-01)	3.317E-01[6](9.734E-02)	**3.194E-03[1](3.714E-03)**
LZ7	4.782E-02[3](2.364E-02)	1.363E-02[2](2.788E-03)	1.808E-01[6](5.384E-04)	8.215E-03[3](7.769E-04)	3.120E-01[7](7.756E-02)	1.443E-01[5](3.930E-02)	**3.274E-03[1](1.400E-03)**
LZ9	1.019E-00[2](1.391E-02)	1.037E-00[3](3.199E-02)	1.110E-00[7](1.114E-01)	6.365E-02[4](1.641E-02)	1.095E-00[4](1.669E-01)	1.122E-01[6](1.759E-02)	**5.366E-03[1](1.611E-03)**
WFG1	9.972E-02[3](1.780E-02)	9.239E-02[4](2.657E-02)	7.887E-01[7](4.710E-01)	1.774E-00[7](1.114E-01)	6.008E-01[6](2.822E-01)	9.064E-02[3](1.495E-02)	9.076E-01[1](5.067E-02)
WFG2	7.691E-02[6](1.511E-02)	7.099E-02[5](1.532E-02)	1.044E-01[7](2.709E-02)	**1.405E-02[1](1.398E-03)**	4.176E-02[3](1.183E-02)	7.018E-02[4](1.249E-02)	2.315E-02[2](3.611E-03)
WFG3	1.154E-01[6](5.885E-03)	1.291E-01[7](2.557E-02)	7.679E-02[4](3.784E-03)	**1.739E-02[1](6.155E-04)**	**2.658E-02[1](2.289E-03)**	7.346E-02[3](5.022E-03)	2.059E-02[2](3.297E-03)
WFG4	7.331E-02[5](1.130E-03)	7.484E-02[6](6.406E-03)	7.975E-02[7](3.784E-03)	8.324E-02[7](4.337E-03)	6.400E-02[4](5.813E-04)	6.355E-02[3](1.346E-03)	5.235E-02[2](8.255E-03)
WFG5	9.071E-02[6](6.034E-03)	9.049E-02[7](7.899E-03)	1.431E-01[7](4.489E-02)	6.331E-02[2](3.303E-03)	5.334E-02[4](7.811E-03)	3.158E-02[3](8.751E-03)	**6.247E-02[1](4.948E-03)**
WFG6	4.048E-02[3](2.225E-03)	3.848E-02[1](4.650E-03)	5.519E-02[7](1.553E-02)	**1.305E-02[1](8.537E-04)**	1.827E-02[3](7.747E-04)	4.607E-02[6](5.963E-03)	3.115E-02[2](1.794E-02)
WFG7	1.404E-01[4](1.408E-02)	1.583E-01[6](2.644E-02)	2.572E-01[7](1.042E-01)	1.693E-02[2](2.417E-03)	1.827E-02[3](7.747E-04)	1.552E-01[5](1.377E-02)	**1.623E-02[1](7.818E-04)**
WFG8	8.843E-02[7](9.919E-03)	8.532E-02[5](1.668E-02)	8.692E-02[6](1.697E-02)	3.891E-02[2](5.397E-03)	1.061E-01[3](9.049E-03)	**1.992E-02[1](1.600E-03)**	**2.020E-02[1](6.239E-03)**
WFG9	—	—	—	3.312E-02[2](3.808E-02)	4.115E-02[4](2.022E-02)	—	3.470E-02[3](1.354E-03)
t/s/≈	31/2/1	28/4/2	27/5/2	27/5/2	27/5/2	24/8/2	

Table 2.6   Overall Performance of 7 Algorithms on the 34 Instances in Terms of *IGD*, *HV* and $I_{\epsilon+}$ Metrics.

Instance	MOEA/D	MOEA/ D-DRA	MOEA/ D-AWA	SMEA	MOCell	SMPSO	DMOEA-εC
Mean Rank	4.7843	3.8431	3.9608	4.2745	4.6667	4.4804	1.9902
Total †/§/ ≈	90/7/5	79/12/11	77/14/11	80/14/8	78/14/10	79/17/6	—

2.6.1.2   *Experimental results on instances with complicated PS shapes (UF and LZ test suites)*

The UF instances come from a set of unconstrained MOP test problems suggested in the CEC2009 contest. The UF and LZ problems involve a strong linkage in variables among the Pareto optimal solutions, thereby posing a great challenge for the MOEAs. Their mathematical descriptions and true PFs can be found in [Zhang *et al.* (2009b)] and [Li and Zhang (2009)].

The experimental results on the bi-objective UF problems in Tables 2.3 to 2.5 demonstrate that DMOEA-εC shows competitive performance on bi-objective UF test problems except UF4, UF5 and UF6 in terms of *IGD*, *HV* and $I_{\epsilon+}$ metrics. As for the UF4 problem, MOCell performs best among all algorithms in terms of three metric values. Nevertheless, even the best performer, MOCell, cannot obtain a good approximation of the PF. However, as for the tri-objective UF test problems, DMOEA-εC performs similarly or outperforms comparison algorithms significantly in terms of the three metrics.

As for the remaining LZ problems, DMOEA-εC outperforms or performs competitively against competitors in terms of *IGD* values. For *HV*, DMOEA-εC shows significant superiority over others on all remaining LZ problems except LZ9. As to $I_{\epsilon+}$, DMOEA-εC has clear advantages over comparison algorithms on all remaining LZ problems except LZ1 and LZ3, on which the performance of DMOEA-εC is slightly worse than that of SMPSO and MOEA/D-AWA, respectively.

In conclusion, the experimental results on UF and LZ test suites indicate that the superiority of DMOEA-εC is significant on tri-objective UF problems and LZ problems, but not significant on bi-objective UF problems, especially on UF4, UF5 and UF6.

2.6.1.3   *Experimental results on instances with complicated PF shapes (WFG test suites)*

In addition to the two above-mentioned suites of test problems, experiments on the WFG test instances are conducted to show the ability of DMOEA-εC on dealing with MOPs with complicated PF shapes. The mathematical descriptions and true PFs of WFG test problems can be found in [Huband *et al.* (2005)].

The experimental results on WFG test suites in Tables 2.3 to 2.5 show that DMOEA-εC performs significantly better than competitors on WFG1, WFG2, WFG5 and WFG7 problems in terms of *IGD*, *HV* and $I_{\epsilon+}$ metric values. Even though DMOEA-εC is not the top among all the algorithms on the rest of the WFG test problems, the rank values of DMOEA-εC on the remaining WFG problems are rightly after the next algorithm. To be specific, SMEA performs better than DMOEA-εC on WFG3, WFG6 and WFG8 problems, significantly in terms of three metrics. Besides, MOCell and SMPSO show their best performance on WFG4 and WFG9 problems, respectively. The reason for the unsatisfactory performance of DMOEA-εC on WFG4 and WFG9 problems could be that DMOEA-εC is not powerful at tackling MOPs with degenerate or deceptive properties. To sum up, DMOEA-εC has shown competitive performance on solving WFG instances with two objectives.

To summarize, as can be seen in Tables 2.3 to 2.5, DMOEA-εC achieves significantly better *IGD* values than MOEA/D, MOEA/DRA, MOEA/D-AWA, SMEA, MOCell and SMPSO in 31, 26, 26, 27, 26 and 25 out of the 34 test instances, respectively. For the *HV*, DMOEA-εC outperforms these competitors significantly in 28, 25, 24, 26, 25 and 30 out of the 34 instances, respectively. For $I_{\epsilon+}$ metric values, DMOEA-εC performs significantly better than these competitors in 31, 28, 27, 27, 27 and 24 out of the 34 instances, respectively. Table 2.6 summarizes these statistical results and reveals the overall rank of the seven algorithms, that is, DMOEA-εC, MOEA/D-DRA, MOEA/D-AWA, SMEA, SMPSO, MOCell and MOEA/D according to the mean ranks. It indicates that DMOEA-εC has the best performance on these continuous test problems in terms of the three metrics. The superiority of DMOEA-εC can be attributed to the efficient information sharing among neighboring subproblems under the ε-constraint framework, the main objective alternation strategy and two matching procedures that strike a balance between convergence and diversity. Note that for a few test problems, some paired algorithms obtain different comparison results with respect to the *IGD* and *HV*, although both indicators measure convergence and diversity. For example, DMOEA-εC has a better *HV* but worse *IGD* than SMPSO on ZDT1 and ZDT2. The reason for this occurrence is the different preferences of the two indicators. The *IGD* is based on uniformly distributed points along the PF and prefers the distribution uniformity of the solution set. However, *HV* is influenced more by boundary solutions and has a bias toward the extensity of the solution set.

For a visual observation, Fig. 2.3 shows the distribution of the final solutions with the minimum *IGD* value within 30 runs found by DMOEA-εC. It is visually evident that for each ZDT and DTLZ instance, the final population obtained by DMOEA-εC can cover the whole PFs very well and spread uniformly. DMOEA-εC shows good convergence and obtains solutions with good diversity on UF1 to UF3 and UF7. For UF4, UF8 to UF10, and LZ problems, final solutions obtained by DMOEA-εC approximate the PFs not very well but spread widely along the PFs. For UF5 and UF6, DMOEA-εC can only find some parts of the PFs. As to the WFG test problems with two objectives, DMOEA-εC achieves good convergence and obtains solutions with good diversity on most of the test instances. In summary, Fig. 2.3 shows that DMOEA-εC can achieve the approximations with both good convergence and diversity for most of the test instances.

2.6.2   *Comparisons on Multi-objective 0/1 Knapsack Problems*

If Ω in (1.1) is a finite set, then Eq. (1.1) is called a combinatorial MOP. The multi-objective 0/1 knapsack problems (MOKPs) are taken as test instances. Extensive experiments are conducted in this part to study and compare DMOEA-εC with MOEA/D on dealing with combinatorial MOPs. A set of nine instances of the MOKPs proposed in [Zitzler and Thiele (1999)] are widely used. MOEA/D outperforms a number of MOEAs without additional local search mechanisms on these test instances. In this chapter, these nine instances are also used for comparing the performances of DMOEA-εC and MOEA/D.

The implementation of DMOEA-εC in terms of operators is exactly the same as MOEA/D [Zhang and Li (2007)]. Specifically, the one-point crossover and standard mutation operator are used to generate a child solution. The greedy repair method proposed by Jaszkiewicz (2002) is adopted after genetic operators. In the greedy repair method, an item with heavy weight in the overfilled knapsacks and little contribution to the single objective function value is more likely to be removed. The initialization of $z_i{}^*$ and $z_i{}^{nad}$ is realized by taking each f_i and $-f_i$ as the objective function and applying the repair method on a randomly generated point, respectively.

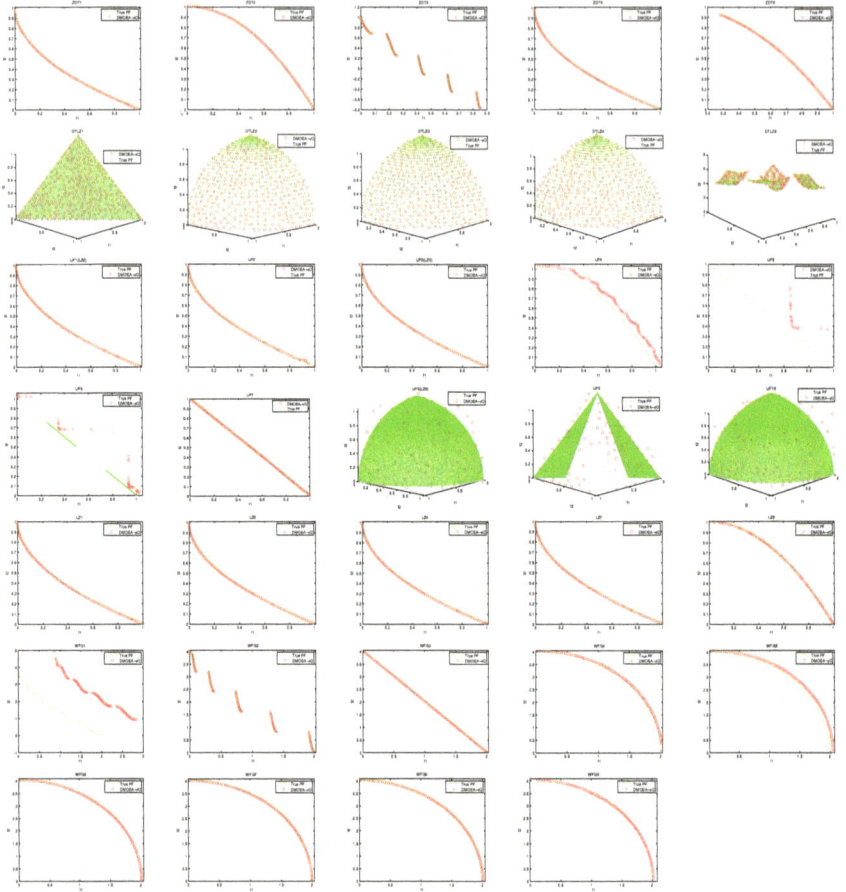

Fig. 2.3   The final populations in the objective space with the minimum *IGD* metric value within 30 runs obtained by DMOEA-εC on 34 continuous problems.

2.6.2.1   *Parameter settings*

Due to the differences in algorithmic frameworks, the population size varies in DMOEA-εC and MOEA/D for each MOKP instance, as illustrated in Table 2.7.[6] The neighborhood size is set as $T = 10$ and the probability of

---

Table 2.7   Parameter Settings of DMOEA-εC and MOEA/D for the Test Instances of 0/1 Knapsack Problems.

Instance				
m	n	N in MOEA/D	N in DMOEA-εC	S
2	250	150	150	150
2	500	200	200	200
2	750	250	250	200
3	250	351	361	200
3	500	351	361	250
3	750	351	361	300
4	250	455	512	300
4	500	455	512	300
4	750	455	512	300

selecting mate solutions from the neighborhood is set as $\delta = 0.9$ for all the instances. The maximal number of solutions replaced by a new solution is set as $n_r = \lfloor 0.01N \rfloor$. The iteration interval of alternating the main objective index IN_m in DMOEA-εC is set as $\lfloor 10\% \cdot (\text{number of iterations}) \rfloor$.

Both DMOEA-εC and MOEA/D stop after $500 \times S$ calls of the repair method. Both of them are independently run 30 times for each test instances on an identical computer.

2.6.2.2   *Experimental results*

The *IGD* and *HV* metrics are used for comparing the performances of different algorithms. In the case where the actual PF is unknown in advance, P^* can be set as an upper approximation of the PF. Jaszkiewicz has produced a good upper approximation to each 0/1 knapsack test instance by solving the linear programming relaxed version of the Tchebycheff formulation of the original multi-objective 0/1 knapsack problem with a number of uniformly distributed weight vectors [Jaszkiewicz (2002)]. The number of the points in the upper approximation is 202 for each of the bi-objective instances, 1326 for the tri-objective instances, and 3276 for the 4-objectives. In our experiments, P^* is set as an upper approximation. The reference points for nine MOKP benchmark problems used in calculations of the *HV* metric values are set to be $\mathbf{r} = \mathbf{0}$.

Table 2.8 gives the means and standard deviations of the *IGD* and *HV* metric values over 30 independent runs of both MOEA/D and DMOEA-εC on the 9 MOKP benchmark instances. The Wilcoxon's rank sum test at

Table 2.8 Statistical Results (Mean (Std. Dev.)) of 2 Algorithms Over 30 Independent Runs on the 9 Instances in Terms of *IGD* and *HV* Metrics.

Instance	250-2	500-2	750-2	250-3	500-3	750-3	250-4	500-4	750-4
IGD									
MOEA/D	5.529E+01†	1.773E+02†	4.201E+02†	2.502E+02†	3.792E+02†	7.526E+02†	2.212E+02†	6.151E+02†	1.055E+03†
	(4.948E-00)	(8.968E-00)	(2.109E+01)	(1.014E+01)	(2.792E+01)	(3.474E+01)	(1.665E+01)	(2.847E+01)	(2.823E+01)
DMOEA-εC	**2.970E+01**	**5.456E+01**	**2.167E+02**	**1.899E+02**	**2.007E+02**	**3.005E+02**	**1.485E+02**	**2.070E+02**	**5.671E+02**
	(1.397E-00)	(3.691E-00)	(1.066E+01)	(9.558E-00)	(9.558E-00)	(9.292E-00)	(1.762E+01)	(1.068E+01)	(1.000E+01)
HV									
MOEA/D	5.239E+07†	6.721E+07†	9.082E+07†	4.837E+11†	1.031E+12†	1.715E+13†	4.833E+15†	3.901E+16†	1.481E+17†
	(3.009E+06)	(5.815E+06)	(1.033E+06)	(3.136E+10)	(1.056E+11)	(2.868E+12)	(1.083E+14)	(1.038E+15)	(3.005E+16)
DMOEA-εC	**4.869E+07**	**6.338E+07**	**1.082E+08**	**8.089E+11**	**4.075E+12**	**3.196E+13**	**6.284E+15**	**8.577E+15**	**4.793E+17**
	(2.967E+06)	(6.235E+06)	(1.140E+07)	(4.204E+11)	(0.747E+11)	(1.962E+12)	(6.186E+13)	(2.350E+15)	(2.648E+16)

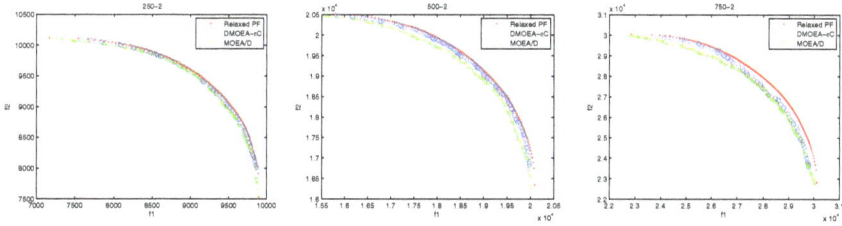

Fig. 2.4 The final populations in the objective space with the minimum *IGD* metric value within 30 runs obtained by MOEA/D and DMOEA-εC on two-objective MOKP test problems.

a 5% significance level is conducted to test the significance of differences between the mean metric values yielded by MOEA/D and the DMOEA-εC. The numbers in parentheses are the standard deviations. †, § and ≈ mean that the performance of DMOEA-εC is better than, worse than, and similar to that of MOEA/D, according to the Wilcoxon's rank sum test, respectively. The bold data in Table 2.8 are the best metric values for each instance. Figure 2.4 plots the distribution of the final approximation with the minimum *IGD* metric value among 30 runs of both MOEA/D and DMOEA-εC for each bi-objective test instance. A relaxed PF represents the upper approximation obtained by Jaszkiewicz.

With the same number of calls of the repair method, it is clear from Table 2.8 that DMOEA-εC performs significantly better than MOEA/D in terms of both *IGD* and *HV* metric values on all the test instances. For example, the average *IGD* values obtained by DMOEA-εC are about 46%, 69% and 48% smaller than those obtained by MOEA/D on instances 250-2, 500-2 and 750-2, respectively. Besides, the larger the number of decision variables and objectives is, the larger differences between DMOEA-εC and MOEA/D are. Figure 2.4 demonstrates that the set of final non-dominated solutions obtained by MOEA/D is dominated by the set obtained by DMOEA-εC on instances 250-2, 500-2 and 750-2. From these figures, it is also clear that the differences in the final approximations between MOEA/D and DMOEA-εC become greater with the increase of decision variables.

In summary, the statistical results on *IGD* and *HV* metric values in Table 2.8 and the distributions of final approximations on bi-objective test problems in Fig. 2.4 confirm the superiority of DMOEA-εC over MOEA/D on solving MOKP benchmark problems.

2.7 Further Discussion

In this section, the parameter analysis and algorithmic behavior of DMOEA-εC are deeply analyzed. First, the influence of the parameter IN_m on the performance of DMOEA-εC is examined. Then, the algorithmic behavior of DMOEA-εC, including effects of both the solution-to-subproblem matching procedure and the subproblem-to-solution matching procedure, as well as the superiority of the farthest-candidate method are further investigated.

2.7.1 *Parameter Sensitivity Analysis of IN_m*

IN_m is a major parameter in DMOEA-εC. It decides how often the algorithm alternates the main objective function. To study how DMOEA-εC is sensitive to this parameter, we take the ZDT1, DTLZ1, DTLZ2, UF1 and WFG2 as examples and test different settings of IN_m in the implementation of DMOEA-εC. Different IN_m values are set as $\lfloor 1\%, 2\%, 5\%, 10\%,$ $20\%, 40\%, 60\%, 80\%, 90\%, 100\% \cdot$ (number of iterations)$\rfloor$. That is to say, the frequency of switching the main objective function becomes smaller and smaller. All the other parameters are kept the same as before. Similarly, 30 independent runs have been conducted for each configuration on these test instances. Figure 2.5 shows the variation of means and standard deviations of IGD and HV metrics across all IN_m values on the selected test problems. As shown in Fig. 2.5, DMOEA-εC performs well with a wide range of IN_m values on UF1 and WFG2. For ZDT1, a large IN_m may be better, while for DTLZ1 and DTLZ2, a small IN_m may be better. Thus it can be claimed that a good setting of IN_m varies with different test instances, such as the settings of IN_m adopted in Section 2.5. Generally, a larger value of IN_m is good for convergence, while a smaller value of IN_m benefits diversity.

Figure 2.5 also reveals that DMOEA-εC with a large IN_m value does not work stably on DTLZ1. A large IN_m value will result in the slow convergence rate and the bad ideal/nadir approximation of the main objective function, which will affect the diversity of the whole population. However, a small value of IN_m means that the solution-to-subproblem matching procedure will be performed more times. This will result in DMOEA-εC consuming more computational resources. Thus, a proper IN_m value strikes a good balance between the performance of DMOEA-εC and the computational cost.

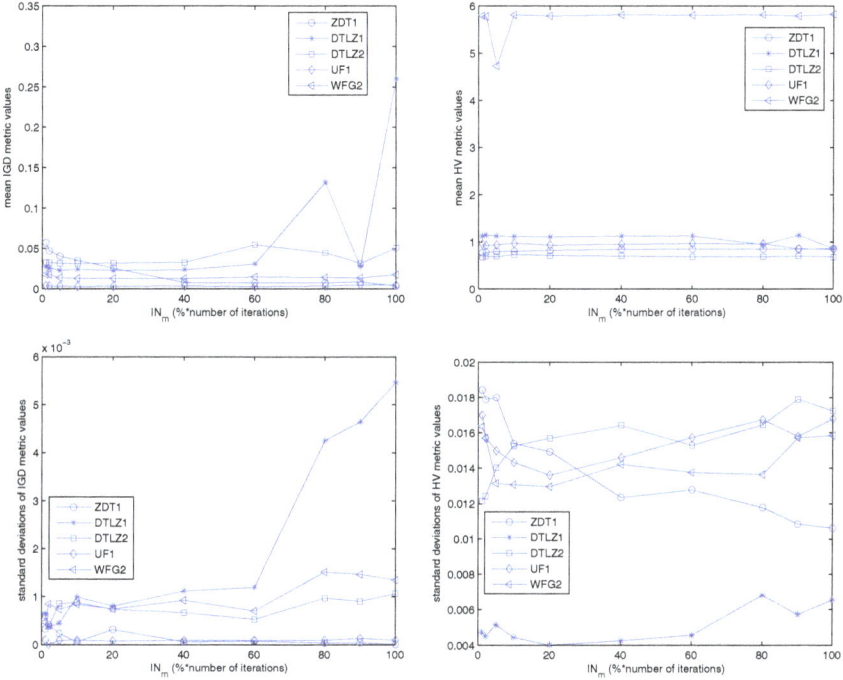

Fig. 2.5   The means and standard deviations of *IGD* and *HV* metric values within 30 runs versus the value of IN_m in DMOEA-εC for ZDT1, DTLZ1, DTLZ2, UF1 and WFG2 test instances.

2.7.2   *Detailed Analysis on Behavior of DMOEA-εC*

This section designs experiments to study the effects of the solution-to-subproblem matching procedure and the subproblem-to-solution matching procedure, and the superiority of the farthest-candidate approach.

2.7.2.1   *Effects of the solution-to-subproblem matching procedure and the subproblem-to-solution matching procedure*

As mentioned above, the solution-to-subproblem matching procedure and the subproblem-to-solution matching procedure strikes a balance between convergence and diversity. Do the above two matching mechanisms indeed play an important role in DMOEA-εC? In order to answer this question, two DMOEA-εC variants, denoted as DMOEA-εC-No_CV and DMOEA-εC-D_No are developed for comparison with the original

DMOEA-εC. Detailed descriptions of the two variants will be given in the following.

Furthermore, why is the distance value from a subproblem to a solution selected as the matching criterion for the solution-to-subproblem matching procedure? Why is the constraint violation value of a solution regarding a subproblem adopted as the matching criterion for the subproblem-to-solution matching procedure? To illustrate the effectiveness of the two criteria, another three variants of DMOEA-εC, including DMOEA-εC-D_D, DMOEA-εC-CV_D and DMOEA-εC-CV_CV, are designed.

DMOEA-εC-No_CV: Different from DMOEA-εC, the solution-to-subproblem matching procedure is removed. And the constraint violation value is still adopted as the matching criterion for the subproblem-to-solution matching procedure.

DMOEA-εC-D_No: In this variant, the subproblem-to-solution matching procedure is removed. And the distance value is still regarded as the matching criterion for the solution-to-subproblem matching procedure.

DMOEA-εC-D_D: In this variant, the distance value is adopted as the matching criterion for the subproblem-to-solution matching procedure.

DMOEA-εC-CV_D: In this variant, the constraint violation value and the distance value are adopted as the matching criterion for the solution-to-subproblem matching procedure and the subproblem-to-solution matching procedure, respectively.

DMOEA-εC-CV_CV: In this variant, the constraint violation value is adopted as the matching criterion for the solution-to-subproblem matching procedure.

All variants are the same as DMOEA-εC except for differences on the two matching procedures. Here we still consider ZDT1, DTLZ1, DTLZ2, UF1 and WFG2 test problems. With the same parameter settings as Section 2.5, the above-mentioned five variants are experimentally compared with DMOEA-εC. The experimental results, in terms of the means and standard deviations of *IGD* and *HV* metric values within 30 independent runs obtained by each algorithm for the selected test instances are all shown in Table 2.9. Similarly, the Wilcoxon's rank sum test at a 5% significance level is conducted to test the significance of differences between the mean metric values yielded by DMOEA-εC and its variants.

Table 2.9 shows that in terms of *IGD* and *HV* metrics, the proposed DMOEA-εC is significantly better than its variants on all selected instances.

Table 2.9 Statistical Results (Mean (Std. Dev.)) of DMOEA-εC and its 5 Variants Over 30 Independent Runs on the 5 Selected Instances in Terms of *IGD* and *HV* Metrics.

Instance	DMOEA-εC-No_CV	DMOEA-εC-D_No	DMOEA-εC-D_D	DMOEA-εC	DMOEA-εC-CV_D	DMOEA-εC-CV_CV
			IGD			
ZDT1	7.400E-03†(6.050E-05)	5.352E-03†(8.450E-05)	1.961E-01†(2.110E-03)	**3.763E-03(6.690E-05)**	2.075E-01†(1.100E-03)	2.115E-01†(4.743E-03)
DTLZ1	4.385E-02†(7.612E-03)	3.530E-02†(2.724E-03)	5.877E-02†(2.702E-01)	**2.476E-02(9.966E-02)**	3.051E-02†(1.972E-03)	6.213E-02†(2.945E-01)
DTLZ2	5.365E-02†(8.610E-03)	4.030E-02†(4.421E-03)	4.755E-02†(6.055E-03)	**3.160E-02(8.819E-04)**	4.395E-02†(2.765E-03)	4.795E-02†(2.205E-01)
UF1	6.500E-03†(7.263E-04)	5.323E-03†(2.810E-04)	5.500E-03†(1.805E-04)	**4.407E-03(9.870E-05)**	5.752E-03†(6.051E-05)	4.845E-03†(4.465E-4)
WFG2	1.655E-02†(2.451E-04)	1.700E-02†(1.300E-04)	1.545E-02†(4.050E-04)	**1.505E-02(8.403E-04)**	1.58E-02†(3.200E-04)	1.610E-02†(1.825E-04)
			HV			
ZDT1	7.476E-01†(4.239E-02)	7.924E-01†(2.480E-02)	6.304E-01†(1.451E-02)	**8.272E-01(1.541E-02)**	6.038E-01†(1.832E-02)	5.965E-01†(2.930E-02)
DTLZ1	1.107E-00†(1.462E-03)	1.118E-00†(1.558E-03)	1.057E-00†(1.517E-03)	**1.307E-00 (4.458E-03)**	1.117E-00†(3.244E-03)	1.047E-00†(1.663E-03)
DTLZ2	6.588E-01†(1.963E-02)	7.294E-01†(1.758E-02)	6.948E-01†(1.537E-02)	**7.435E-01 (1.528E-02)**	7.094E-01†(1.704E-02)	7.267E-01†(1.694E-02)
UF1	8.637E-01†(1.436E-02)	8.700E-01†(3.891E-02)	8.511E-01†(6.481E-02)	**8.741E-01(1.434E-02)**	8.471E-01†(8.457E-02)	8.628E-01†(6.590E-02)
WFG2	5.926E-00†(1.199E-02)	5.981E-00†(1.693E-02)	5.862E-00†(1.054E-02)	**6.076E-00(1.310E-02)**	5.95E-00†(2.479E-02)	5.905E-00†(5.793E-02)

The effectiveness of the solution-to-subproblem matching procedure using the distance value as the matching criterion, and the subproblem-to-solution matching procedure adopting the constraint violation value as the matching criterion is confirmed experimentally.

2.7.2.2 *Superiority of the farthest-candidate method*

In order to further investigate the superiority of the farthest-candidate method when pruning EP, we compare it with the crowding distance-based mechanism used in NSGA-II [Deb *et al.* (2002a)]. Thus, we develop a DMOEA-εC variant, denoted as the DMOEA-εC-CD, by replacing the farthest-candidate method in DMOEA-εC with the crowding distance-based mechanism.

Take ZDT1, DTLZ1, DTLZ2, UF1 and WFG2 test instances as examples. DMOEA-εC-CD has been experimentally compared with DMOEA-εC with the same parameter settings as Section 2.5. The experimental results, in terms of the means and standard deviations of IGD and HV metric values of the final solutions within 30 independent runs obtained by each algorithm for the selected test instances are all shown in Table 2.10. Similarly, the Wilcoxon's rank sum test at a 5% significance level is conducted to test the significance of differences between the mean metric values yielded by DMOEA-εC and its variant DMOEA-εC-CD.

Table 2.10 demonstrates the superiority of DMOEA-εC over DMOEA-εC-CD in terms of both IGD and HV metric values on five selected test

Table 2.10   Statistical Results (Mean (Std. Dev.)) of DMOEA-εC and its Variant DMOEA-εC-CD Over 30 Independent Runs on 5 Selected Test Instances in Terms of IGD and HV Metrics.

	IGD		*HV*	
	DMOEA-εC	DMOEA-εC-CD	DMOEA-εC	DMOEA-εC-CD
ZDT1	**3.763E-03**	8.250E-03[†]	**8.272E-01**	8.072E-01[†]
	(6.690E-05)	(2.881E-05)	(1.541E-02)	(8.840E-02)
DTLZ1	**2.476E-02**	3.385E-02[†]	**1.307E-00**	1.126E-00[†]
	(9.966E-04)	(1.821E-03)	(4.458E-03)	(3.549E-03)
DTLZ2	**3.160E-02**	3.595E-02[†]	**7.435E-01**	7.067E-01[†]
	(8.819E-04)	(6.841E-04)	(1.528E-02)	(1.280E-02)
UF1	**4.407E-03**	6.500E-03[†]	**8.741E-01**	8.554E-01[†]
	(9.870E-05)	(2.451e-04)	(1.434E-02)	(1.053E-02)
WFG2	**1.505E-02**	1.62E-02[†]	**6.076E-00**	5.992E-00[†]
	(8.403E-04)	(3.216E-04)	(1.310E-02)	(3.793E-02)

problems. This also shows the rationality and superiority of the farthest-candidate method when pruning the external archive population.

2.8 Conclusion

Decomposition and the ε-constraint method are two important strategies in the field of multi-objective optimization. This chapter has reformulated MOPs by incorporating the ε-constraint method into the decomposition strategy and proposed the decomposition-based multi-objective evolutionary algorithm with the ε-constraint framework (DMOEA-εC) to deal with MOPs. DMOEA-εC explicitly decomposes an MOP into a series of scalar-constrained optimization subproblems by selecting one of the objectives as the main objective function and assigning each subproblem with an upper bound vector ε. Then these subproblems are optimized simultaneously by evolving a population of solutions. At each generation, each individual solution in the population is associated with a subproblem. The neighborhood relations among these subproblems are defined based on the Euclidean distance between their upper bound vectors. And the assumption that optimal solutions of two neighboring subproblems should be very similar is still valid. Besides, a main objective alternation strategy, a solution-to-subproblem matching procedure, and a subproblem-to-solution matching procedure are proposed to strike a balance between convergence and diversity.

DMOEA-εC has been compared with six state-of-the-art MOEAs, i.e., MOEA/D [Zhang and Li (2007)], MOEA/D-DRA [Zhang *et al.* (2009a)], MOEA/D-AWA [Qi *et al.* (2014)], SMEA [Zhang *et al.* (2016)], MOCell [Nebro *et al.* (2009a)] and SMPOS [Nebro *et al.* (2009b)] on 34 continuous test instances and 8 MOKPs test problems. A systematical experimental study has demonstrated that DMOEA-εC outperforms or performs competitively against other algorithms on the majority of the test instances. The sensitivity of the parameter IN_m in DMOEA-εC has been experimentally investigated. Moreover, the algorithmic behavior of DMOEA-εC including the effects of both the solution-to-subproblem matching procedure and the subproblem-to-solution matching procedure as well as the superiority of the farthest-candidate method have been further analyzed. All these experimental results confirm that DMOEA-εC can deal with the majority of the continuous benchmark problems and the MOKP test problems successfully.

Future research work includes investigations of adopting alternative methods to solve each constrained subproblem, employing more effective methods for estimating the nadir point and proposing an adjustment strategy for upper bound vectors to further improve the uniformity of the final population. Besides, based on our previous research works [Chen *et al.* (2009b); Xin *et al.* (2012)], the hybridization of different search operators in DMOEA-εC is also worthwhile to be studied.

Chapter 3

Decomposition-based Many-objective Evolutionary Algorithm with the ε-Constraint Framework

Decomposition-based MOEAs have emerged as promising approaches for tackling MOPs. The decomposition-based method decomposes an MOP into a set of subproblems and optimizes them in a collaborative manner. The decomposition concept is so general that either aggregation functions [Chen *et al.* (2017); Zhang and Li (2007)] or simple MOPs [Liu *et al.* (2014)] can be used to form subproblems. The multi-objective evolutionary algorithm based on decomposition (MOEA/D) [Zhang and Li (2007)] is a representative of this sort. MOEA/D converts an MOP into a set of scalar subproblems by using aggregated functions and optimizes them concurrently. Commonly used aggregated functions in MOEA/D include the weighting method, the Tchehycheff method, and the penalty-based boundary intersection (PBI) method. Apart from the abovementioned aggregation methods, the ε-constraint method is also a basic generation method in mathematical programming and is often used as an element of more developed methods [Miettinen (1999)]. The decomposition-based multi-objective evolutionary algorithm with the ε-constraint framework (DMOEA-εC) is a newly proposed MOEA in recent research [Chen *et al.* (2017)]. DMOEA-εC explicitly decomposes an MOP into a series of scalar-constrained optimization subproblems by associating each subproblem with an upper bound vector. These subproblems are optimized collaboratively by an evolutionary algorithm based on the feasibility rule [Deb (2000)]. In decomposition-based MOEAs, the diversity of a population is implicitly maintained by a set of reference vectors and all subproblems are optimized simultaneously by using information from neighboring subproblems. So far, the effectiveness of MOEA/D on MaOPs has been investigated [Branke *et al.* (2001); Hadka and Reed (2012); Thiele *et al.* (2014)], but similar research for DMOEA-εC is still lacking.

In this chapter, an improved version of DMOEA-εC for both MOPs and MaOPs, named as DMaOEA-εC, is proposed. DMaOEA-εC explicitly decomposes an MOP into a series of scalar-constrained optimization subproblems and optimizes each subproblem based on a newly proposed two-side update rule. The proposed two-side update rule maintains both feasible and infeasible solutions for each subproblem, which is more effective than the feasible rule. The main objective alternation strategy and the solution-to-subproblem matching procedure are still kept in DMaOEA-εC. Next, with an aim of making DMaOEA-εC applicable to MaOPs, a two-stage upper bound vectors generation procedure is presented to generate widely spread upper bound vectors in a high-dimensional objective space. Besides, in order to remedy the diversity loss of a population in decomposition-based approaches, a distance-based global replacement strategy is put forward to compare a newly generated solution with all subproblems in a distance-based order. Furthermore, a boundary point maintenance mechanism is introduced to reduce the possibility of losing boundary solutions and thus achieve a good spread of solutions over the PF.

3.1   Framework of DMaOEA-εC

3.1.1   *Strengths and Drawbacks of DMOEA-εC*

DMOEA-εC converts an MOP into a series of scalar-constrained subproblems and then optimizes them collaboratively by using the neighbour information. It has shown obvious advantages over several state-of-the-art MOEAs on continuous and discrete MOPs. However, as the number of upper bound vectors increase exponentially with the number of objectives, it is unadvisable to apply DMOEA-εC to handle MaOPs directly. Besides, the diversity of a population is maintained by a set of predefined well distributed upper bound vectors, which is not enough in many-objective optimization. DMOEA-εC struggles to maintain the diversity in many-objective optimization, and usually fails to achieve a good coverage of the PF. Thus, another drawback of DMOEA-εC is the lack of improved diversity maintenance mechanisms in the many-objective optimization process. Furthermore, since the feasibility rule adopted in DMOEA-εC is straightforward but simple, the performance of DMOEA-εC can be further improved.

3.1.2 Framework of DMOEA-εC

DMaOEA-εC converts an MOP into N scalar-constrained subproblems and optimizes them simultaneously in a single run. Let $\{\varepsilon^1,\varepsilon^2,\ldots,\varepsilon^N\}$ be a set of evenly spread upper bound vectors, and the neighborhood of upper bound vector ε^i is defined as a set of its several closest upper bound vectors in $\{\varepsilon^1,\varepsilon^2\ldots,\varepsilon^N\}$. The neighbor of the ith subproblem consists of all subproblems with the upper bound vectors from the neighborhood of ε^i and is denoted as $B(i)$. When optimizing each subproblem, a newly proposed two-side update rule is adopted to handle constrained subproblems.

The following notations will be used in the description of DMaOEA-εC.

- N: Number of upper bound vectors, which is the same as the population size;
- T: Neighborhood size;
- δ: Probability of selecting mating solutions from its neighborhood;
- N_r: Maximum number of replacements;
- IN_m: Iteration interval of alternating the main objective;
- S: Maximum size of the external archive population EP; and
- NFE: Maximum number of function evaluations.

The algorithmic description of DMaOEA-εC is presented in **Algorithm 6**, in which *rand* is a random value in $[0,1]$.

3.1.3 Two-stage Upper Bound Vectors Generation Process

According to the systematic approach developed in [Chen *et al.* (2017)], $N = (q+1)^{m-1}$ points will be generated for an m-objective problem with q divisions along each objective axis. In order to have intermediate upper bound vectors, we should set $q \geq 2$. However, in a high-dimensional objective space, there will be a large amount of upper bound vectors even if $q = 1$. For example, for a 15-objective case, $q = 1$ will result in $2^{14} = 16,384$ upper bound vectors. This obviously aggravates the computational burden. If we simply remedy this issue by lowering q (e.g., set $q = 1$), it will make all upper bound vectors lie along the boundary of the simplex. This is apparently harmful to the population diversity. To avoid such a situation, we present a two-stage upper bound vectors generation process, as described in **Algorithm 7**.

Algorithm 6 Framework of DMaOEA-εC

Require: An MOP or MaOP, related parameters.
Ensure: An external archive population EP.
1: Initialize N evenly spread upper bound vectors using the two-stage upper bound vectors generation process (**Algorithm 7**).
2: Initialize the evolving population $\mathbf{P} = \{\mathbf{x}^1, \ldots, \mathbf{x}^N\}$ and set $\mathbf{FV}^i = \mathbf{F}(\mathbf{x}^i)$ randomly; extract non-dominated individuals from P and denote the set of them as EP; initialize $\mathbf{z}^*$ and $\mathbf{z}^{\mathbf{nad}}$; set $gen = 0, n = N$.
3: Use the solution-to-subproblem matching procedure (**Algorithm 8**) to match solutions with subproblems.
4: **for** $i = 1$ to N **do**
5:     Set the neighborhood of the ith subproblem $B(i)$.
6: **end for**
7: **while** $n \leq NFE$ **do**
8:     **if** gen is a multiple of IN_m **then**
9:         Alternate the main objective.
10:         Use the solution-to-subproblem matching procedure (**Algorithm 8**) to match solutions with subproblems.
11:     **end if**
12:     **for** $i \in I$ **do**
13:         $P = \begin{cases} B(i), & \text{if } rand < \delta \\ \{1, 2, \ldots, N\}, & otherwise \end{cases}$
14:         Reproduction: Select parent individuals from P randomly and apply a certain reproduction operator to generate a new solution $\mathbf{y}$.
15:         $n = n + 1$.
16:         Repair: If $\mathbf{y}$ is infeasible, repair it.
17:         Update the approximated ideal point $\mathbf{z}^*$.
18:         Update the external archive EP and prune it by using the crowding distance.
19:         Update the approximated nadir point $\mathbf{z}^{\mathbf{nad}}$.
20:         Apply the boundary points maintenance mechanism.
21:         Compute the distance value from $\mathbf{y}$ to all upper bound vectors ε^i, i.e., $d_i^y, i = 1, 2, \ldots, N$, and sort the distance value in ascending order, i.e., $d_{k_1}^y \leq d_{k_2}^y \leq \cdots \leq d_{k_N}^y$.// $d_i^y = \sum_{j=1}^{m} |f_j^y - \varepsilon_j^i|$.
22:         Set $j = 1, nr\_count = 0$.
23:         **while** $j \leq N$ and $nr\_count \leq nr$ **do**
24:             Compare $\mathbf{y}$ with the solution of the subproblem k_j by using the proposed two-side update rule, and if $\mathbf{y}$ can replace the solution of subproblem k_j then $nr\_count = nr\_count + 1$.
25:             $j = j + 1$.
26:         **end while**
27:     **end for**
28:     $gen = gen + 1$.
29: **end while**

Algorithm 7 Two-stage Upper Bound Vectors Generation Procedure

---

Require: The number of objectives m, the number of divisions along each objective axis for the alternative set of upper bound vectors q, and the population size N.

Ensure: A set of evenly spread upper bound vectors A.

1: Generate the set of upper bound vectors that $(0,0,\ldots,0)$, $(1,1,\ldots,1)$, and permutations of $(1,0,\ldots,0)$ and denote it as $A = \{\mathbf{a}^1, \mathbf{a}^2 \ldots, \mathbf{a}^{N_1}\}$ $(N_1 = m+1)$.

2: Use the generation proposed in [Chen *et al.* (2017)] to generate $N_2 = (q+1)^{m-1}$ upper bound vectors $B = \{\mathbf{b}^1, \mathbf{b}^2 \ldots, \mathbf{b}^{N_2}\}$, where $N_1 \ll N_2, N_1 + N_2 > N$.

3: **while** $|A| < N$ **do**

4:     Select the upper bound vector $\mathbf{b} \in B$, which is farthest from the selected upper bound vectors A.

5:     $A = A \cup \{\mathbf{b}\}$, $B = B\backslash\{\mathbf{b}\}$.

6: **end while**

Algorithm 8 Solution-to-Subproblem Matching

Require: N solutions $(\mathbf{x}^1, \mathbf{FV}^1), \ldots, (\mathbf{x}^N, \mathbf{FV}^N)$ and N subproblems with upper bound vectors $\varepsilon^1, \varepsilon^2 \ldots, \varepsilon^N$.

Ensure: Matched pairs $(\mathbf{x}^k, \mathbf{FV}^k) \sim \varepsilon^l (k, l = 1, \ldots, N)$.

1: Initialize $S = \{1, 2, \ldots, N\}$.

2: **while** S is nonempty **do**

3:     Randomly select an upper bound vector $\varepsilon^l, l \in S$.

4:     **for** $i = 1$ to N **do**

5:         $d_i^l = \sum_{j=1, j\neq s}^m |f_j^i - \varepsilon_j^l|$.

6:     **end for**

7:     $k = argmin_{i \in \{1,2,\ldots,N\}} d_i^l$.

8:     $\mathbf{FV}^k = \mathbf{Inf}$; $S = S\backslash\{l\}$.

9: **end while**

At first, the set of upper bound vectors, which are $(0,0,\ldots,0)$, $(1,1,\ldots,1)$, and permutations of $(1,0,\ldots,0)$, is generated and denoted as $A = \{\mathbf{a}^1, \mathbf{a}^2 \ldots, \mathbf{a}^{N_1}\}$ $(N_1 = m+1)$.[1] Then, an alternative set of upper bound vectors, denoted as $B = \{\mathbf{b}^1, \mathbf{b}^2 \ldots, \mathbf{b}^{N_2}\}$, is initialized according

---

[1] Optimal solutions of subproblems with these upper bound vectors are boundary points whose exact definition will be given in the following.

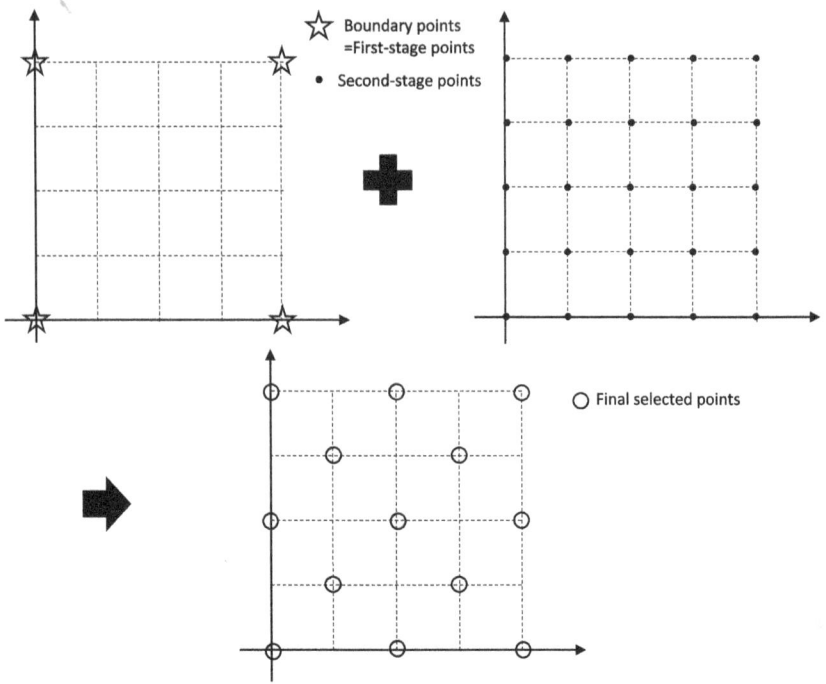

Fig. 3.1   Example of the two-stage upper bound vectors generation procedure for a tri-objective problem with $q = 4$, and $N = 13$ upper bound vectors are selected eventually.

to the method adopted in [Chen *et al.* (2017)] using a q setting. Next the upper bound vector in B that has the largest distance value to the set A is selected and added into A. The distance between an upper bound vector $\mathbf{b}^i$ and a set of upper bound vectors $A = \{\mathbf{a}^1, \ldots, \mathbf{a}^{|A|}\}$ is defined as $d(\mathbf{b}^i, A) = \min_{j=1,\ldots,|A|} \{\|\mathbf{b}^i - \mathbf{a}^j\|\}$. Set A is updated and this process continues until the number of upper bound vectors in A reaches the predefined population size N. Figure 3.1 presents a simple example to illustrate the two-stage upper bound vectors generation procedure. To be specific, this procedure results in $N = 13$ upper bound vectors for a tri-objective optimization problem ($m = 3$) by using a spacing of $\delta = 1/4$ ($q = 4$).

3.1.4   *Solution-to-Subproblem Matching Procedure*

Similar to DMOEA-εC, after the main objective alternation strategy is utilized, a solution that is good for the current subproblem will no longer perform well since the objective function of this subproblem has been changed. Thus, a solution-to-subproblem matching procedure is proposed. In this matching procedure, the solution that has the minimum distance value to certain subproblems among N current solutions is matched with this subproblem. The distance value from solution $(\mathbf{x}^i, \mathbf{FV}^i)$ to the subproblem with ε^l is defined as $d_i^l = \sum_{j=1, j \neq s}^{m} |f_j^i - \varepsilon_j^l|$. This matching procedure will be used after the main objective alternation procedure. It can place the nearest solution to each subproblem at a large degree, which is beneficial to diversity.

3.1.5 *Boundary Points Maintenance Mechanism*

If a solution is generated by simultaneously minimizing $k(k < m)$ objectives of an m-objective problem, it is referred as a corner solution [Wang and Yao (2013)]. Corner solutions play an important role in MOPs, especially in MaOPs since they have a good ability in representing the bound of each objective on the PF and are definitely non-dominated individuals. For a bi-objective problem, it is straightforward to visualize the corner solutions corresponding to the minima of both objectives. However, for a generic m-objective problem, k can take values from 1 to $m - 1$, hence there are many possible corner solutions to an m-objective optimization problem. The number of corner solutions corresponding to all possible sets of objectives minimizes, of which results in a single optimal solution [Wang and Yao (2013)].[2]

Here we only take the cases that $k = 1$ and $k = m - 1$ into consideration. Solutions that are optimal to each objective f_i and the L_2 norm of the rest of the objectives except the current one f_i $(i = 1, \ldots, m)$ are termed as boundary points [Bechikh (2013)]. Prioritizing these boundary individuals as described in line 20 of **Algorithm 6** not only reduces the possibility of losing boundary solutions, but also enhances the ability of covering the PF as widely as possible. In this way, boundary solutions are promoted over generations in order to eventually capture corners of the PF. Out of the above reasons, it is reasonable to first select boundary individuals and put them into the current population. Thus, the boundary points maintenance mechanism is suggested by taking advantage of boundary solutions to achieve a good spread of solutions over the PF.

---

[2]The case where minimizing all objectives simultaneously results in a single optimal solution has been omitted, because in that case the problem would be reduced to a SOP.

3.1.6   *Distance-based Global Replacement Strategy*

The diversity of a population in DMOEA-εC is maintained by a set of uniformly spread upper bound vectors. However, solutions do not spread as well as the upper bound vectors, which results in bad diversity of obtained solutions along the PF. This situation could be attributed to the fact that the solution, which is distant to certain upper bound vectors, could also achieve relatively good objective value for a corresponding subproblem. Besides, the evolutionary process would be misled if solutions are selected only depending on aggregation function values. It is worth noting that solutions are usually far away from the PF and the misleading selection is more likely to occur in the early stage of evolution, which pulls the search process to only a part of the PF. What's more, the above-mentioned issues would be aggravated in a high-dimensional objective space due to the sparse distribution of solutions.

Given all of the above, we are motivated to consider not only the aggregation function value of a solution but also the distance between the solution and corresponding upper bound vector in DMaOEA-εC. This practice is expected to force each solution to stay close to an upper bound vector and explicitly maintain the desired distribution of solutions in the evolutionary process, which leads to a better balance between convergence and diversity in many-objective optimization. The proposed global distance-based replacement strategy gives priority to diversity over convergence, which is preferable in solving MaOPs. Besides, it should be noted that the concept of the neighbor is only used in the mating process of DMaOEA-εC.

Details of the proposed distance-based global replacement strategy are given in lines 21–26 of **Algorithm 6**. To be specific, once a new solution **y** is produced, the distance value between **y** and each upper bound vector ε_i is computed as $d_i^y = \sum_{j=1}^m |f_j^y - \varepsilon_j^i| (i = 1, \ldots, N)$. Then these distance values are sorted in an ascending order. Suppose the N distance values corresponding to N subproblems are arranged as $d_{k_1}^y \leq d_{k_2}^y \leq \cdots \leq d_{k_N}^y$. Then, solution **y** is compared with solutions $x^{k_1}, \ldots, x^{k_N}$ until the maximum number of replacement n_r is reached or all the solutions have been compared with **y**.

3.1.7   *Two-side Update Rule*

The feasibility rule adopted in DMOEA-εC is straightforward but simple for constrained problems. Here, a new two-side update rule, which maintains both feasible and infeasible solutions for each subproblem is presented. To be specific, when a new solution **y** and an old solution $\mathbf{x}^i$ corresponding to the ith subproblem are compared, the following criteria are followed:

Case 1: **y** is feasible for the ith subproblem:
If there is a feasible solution for ith subproblem, i.e., $\mathbf{x}^i$, the one having better objective function value if preferred; otherwise, **y** is definitely accepted as the feasible solution for the ith subproblem.

Case 2: **y** is infeasible for the ith subproblem:
If there is an infeasible solution for the ith subproblem, i.e., $\mathbf{x}^i$, the one having a smaller constraint violation is preferred; otherwise, **y** is definitely accepted as the infeasible solution for the ith subproblem.

It should be noted that the newly updated solutions are regarded as the active solution for each subproblem. The set of all active solutions is regarded as the mating pool for generating new solutions.

3.1.8 *Main Differences Between DMaOEA-εC and DMOEA-εC*

Both DMaOEA-εC and DMOEA-εC incorporate the ε-constraint method into the decomposition strategy and solve an MOP via optimizing a series of scalar-constrained subproblems collaboratively using the neighbor information. However, they adopt different update rules to tackle each constrained subproblem. The concept of a neighbor is abandoned in the replacement phase and is only used in the mating process in DMaOEA-εC. Besides, DMaOE-εC compares the new solution with all subproblems in a distance-based order instead of comparing with neighboring subproblems in a random order as in DMOEA-εC. What's more, a two-stage upper bound vectors generation process and a boundary points maintenance mechanism are embedded in DMaOEA-εC to make it applicable in the many-objective optimization.

3.2 Experimental Design

3.2.1 *Multi-objective and Many-objective Continuous Test Instances*

The ZDT test instances, UF test suites [Zhang *et al.* (2009b)] (which are part of the CEC2009 MOP test instances), LZ test suites [Li and

Zhang (2009)] (with complicated PS shapes), and bi-objective WFG test suites [Huband *et al.* (2005)] (with complicated PF shapes), are adopted for comparing DMaOEA-εC with five other MOEAs on MOPs.

The DTLZ test instances [Deb *et al.* (2002b)] and WFG test suites with 3/5/8/10 objectives are adopted for comparing the performance of DMaOEA-εC with its competitors on MaOPs.

3.2.2 *Performance Metrics*

Two commonly used performance metrics, i.e., inverted generational distance (IGD) [Zhou *et al.* (2005)] and hypervolume (HV) [Zitzler and Thiele (1999)] are employed to evaluate the performance of all compared algorithms.

3.2.3 *MOEAs for Comparison*

As to MOPs, five representive MOEAs are considered as competitive candidates, including NSGA-II [Deb *et al.* (2002a)], IBEA [Zitzler and Kunzli (2004)], MOEA/D [Zhang and Li (2007)], MOEA/IGD-NS [Tian *et al.* (2016)] and DMOEA-εC [Chen *et al.* (2017)]. NSGA-II, IBEA and MOEA/D are representative Pareto-, indicator- and decomposition-based MOEAs, respectively. MOEA/IGD-NS is a newly proposed decomposition-based MOEA and performs better than a number of popular algorithms. DMOEA-εC is the original of the proposed DMaOEA-εC and has shown its advantages over competitive approaches.

As to MaOPs, four state-of-the-art MOEAs designed for MaOPs are considered as competitive candidates, including HypE [Bader and Zitzler (2011)], NSGA-III [Deb and Jain (2014)], MOEADD [Li *et al.* (2015a)] and Two_Arch2 [Wang *et al.* (2015)]. HypE is an indicator-based MOEA and estimates the HV contribution based on Monte Carlo sampling to make HV-based selection viable. Two_Arch2 maintains two archives for measuring convergence and diversity separately during the evolutionary search process. MOEADD and NSGA-III both combine dominance- and decomposition-based approaches and perform better than a number of popular algorithms.

3.2.4 *Parameter Settings*

For a fair comparison, the choice of parameters are the same as the comparison algorithms. Specifically, the population size is set to $N = 100$ for the ZDT and bi-objective WFG problems. As to the UF problems, population size is set to $N = 600$ for the bi-objective and $N = 1000$ for the tri-objective. Since 4 out of the 9 LZ test problems are included in the UF test suites, the population size is set to $N = 300$ for the remaining LZ problems. Besides, the population size is set to $N = 351/212/156/276$ for the DTLZ problems with $3/5/8/10$ objectives. For WFG problems with $3/5/8/10$ objectives, the population size is set to $N = 200$.

Besides, an external archive population with the size of $S = \lfloor 1.5N \rfloor$ is added to the comparison algorithms. The DE operator and Gaussian mutation are used in solving ZDT, DTLZ and WFG test problems. The DE operator and polynomial mutation are adopted in solving UF and LZ test problems [Chen *et al.* (2017)]. Moreover, control parameters for these reproduction operators are the same as those claimed in comparative MOEAs.[3] All compared algorithms stop when the number of function evaluations reaches the maximum number. For a fair comparison, in accordance with the parameter settings in comparison algorithms, the maximum number of function evaluations is set to 50,000 for the ZDT problems, 300,000 for the UF problems, 150,000 for the remaining LZ problems, and 45,000 for the bi-objective WFG problems. As to MaOPs, Table 3.1 shows the maximum number of function evaluations for different DTLZ test instances. Besides, the maximum number of function evaluations for all WFG test instances is set as 50,000. Finally, each algorithm is executed 30 times independently on each instance.

For DMaOEAεC, the neighborhood size is set as $T = \lfloor 0.1N \rfloor$; the probability of selecting mate solutions from the neighborhood is given as $\delta = 0.9$; the maximal number of replacement is set to $n_r = \lfloor 0.01N \rfloor$ [Chen *et al.* (2017)]. According to [Bader and Zitzler (2011)], the number of Monte Carlo sampling is set as 10,000.

Table 3.1   The Maximum Number of Function Evaluations for DTLZ Instances.

Instance	$m = 3$	$m = 5$	$m = 8$	$m = 10$
DTLZ1	250	600	750	1000
DTLZ2	250	350	500	750
DTLZ3	250	1000	1000	1500
DTLZ4	250	1000	1250	2000

---

[3]CR is changed to 0.9 only for ZDT problems.

3.3   Numerical Results on Multi- and Many-objective Test Instances

This section is devoted to the experimental design for investigating the performance of DMaOEA-εC on multi- and many-objective test instances.

3.3.1   *Comparisons on MOPs*

This part of the experiment is designed to study the effectiveness of DMaOEAεC on MOPs. The means and standard deviations of *IGD* and *HV* metric values over 30 independent runs of each algorithm on 29 MOP instances are shown in Tables 3.2 and 3.3, respectively. The mean *HV* (*IGD*) values for each instance are sorted in descending (ascending) order, and the numbers in the square brackets are their ranks. The Wilcoxon's rank sum test at a 5% significance level is conducted to test the significance of differences between the mean metric values yielded by DMaOEA-εC and comparison algorithms. The numbers in parentheses are the standard deviations, and the bold data in the table are the best mean metric values for each instance.

As can be seen from Table 3.2, in terms of *IGD* metric values, DMaOEA-εC outperforms all comparison algorithm on all ZDT instances except ZDT4, on which DMaOEA-εC and MOEA/D show similar performance. DMaOEA-εC exhibits the best performance on the majority of UF problems. However, as to UF4, UF6 and UF8, DMaOEA-εC shows slightly worse performance compared with other algorithms. Besides, NSGA-II performs the best on UF4, and MOEA/IGD-NS outperforms competitors on UF6 and UF8. For all LZ test problems, DMaOEA-εC outperforms or performs competitively against competitors. DMaOEA-εC has clear advantages over comparison algorithms on all bi-objective WFG problems except WFG1 and WFG4, on which the performance of DMOEA-εC is slightly worse than that of competitors. To summarize, as can be seen in Table 3.2, DMaOEA-εC achieves significantly better *IGD* values than NSGA-II, IBEA, MOEA/D, MOEA/IGD-NS and DMOEA-εC in 23, 26, 26, 22 and 16 out of the 29 test instances, respectively.

In terms of the *HV* metrics as shown in Table 3.3, DMaOEA-εC exhibits a significant advantage over competitors on all ZDT instances. Similar to the results in terms of *IGD* measures, DMaOEA-εC performs best on all UF instances except UF4 and UF6, on which DMaOEA-εC performs similarly with comparison algorithms. DMaOEA-εC exhibits

Table 3.2   Statistical Results of 7 Algorithms over 30 Independent Runs on the 29 MOP Instances in Terms of *IGD* Metrics.

Instance	NSGA-II	IBEA	MOEA/D
ZDT1	4.606E-03†[5](3.817E-06)	4.661E-03†[6](2.226E-06)	3.800E-03†[3](0.000E-07)
ZDT2	4.739E-03†[5](3.484E-06)	9.690E-03†[6](6.981E-07)	4.000E-03†[4](1.830E-05)
ZDT3	3.482E-02†[5](1.432E-03)	1.029E-01†[6](9.598E-04)	1.787E-02†[3](1.255E-02)
ZDT4	4.739E-03†[5](6.113E-07)	6.228E-02†[6](9.0256E-03)	**3.890E-03≈[1](7.588E-05)**
ZDT6	3.732E-03†[5](2.626E-06)	5.064E-03†[6](1.475E-06)	3.330E-02†[4](0.000E-07)
UF1(LZ2)	9.052E-02†[5](1.532E-04)	1.007E-01†[6](6.479E-05)	9.397E-03†[3](1.623E-03)
UF2(LZ5)	2.301E-02†[4](1.581E-05)	2.738E-02†[5](2.246E-05)	1.436E-02†[3](3.333E-03)
UF3(LZ8)	1.578E-01†[4](2.241E-03)	2.023E-01†[6](1.299E-03)	2.107E-02†[3](1.990E-02)
UF4	**4.021E-02§[1](1.986E-07)**	4.083E-02§[2](4.844E-07)	6.907E-02§[4](7.412E-03)
UF5	**1.920E-01§[1](1.463E-03)**	2.479E-01†[4](2.756E-03)	3.355E-01†[5](1.409E-01)
UF6	1.653E-01§[2](4.876E-03)	2.102E-01§[3](5.514E-03)	3.096E-01†[6](2.215E-01)
UF7	4.972E-02§[2](4.672E-03)	4.088E-02†[3](4.041E-04)	7.633E-03†[3](3.333E-03)
UF8(LZ6)	3.359E-02§[2](2.200E-03)	4.774E-01†[6](3.886E-04)	7.988E-02†[5](1.389E-02)
UF9	2.241E-02§[2](2.770E-04)	7.038E-02†[5](4.280E-03)	1.353E-01†[6](4.758E-02)
UF10	1.203E-01†[4](4.701E-03)	4.477E-01†[5](1.556E-03)	4.490E-01†[6](7.552E-02)
LZ1	4.049E-02†[6](9.501E-05)	5.138E-03†[4](4.364E-05)	1.290E-03†[2](3.724E-05)
LZ3	5.793E-02†[5](2.43E-04)	4.575E-02†[4](1.937E-04)	6.122E-03†[3](3.810E-03)
LZ4	6.939E-02†[5](2.609E-03)	4.932E-02†[4](7.128E-05)	7.265E-03†[3](3.758E-03)
LZ7	1.457E-01†[4](1.432E-03)	2.012E-01†[6](4.564E-03)	3.056E-02†[3](4.478E-02)
LZ9	1.176E-01†[4](1.213E-03)	2.134E-01†[6](6.208E-03)	9.770E-03†[3](4.729E-03)
WFG1.2	2.392E-01§[2](4.566E-03)	**8.974E-02§[1](4.738E-04)**	1.202E-00†[6](2.487E-02)
WFG2.2	1.846E-02†[3](2.240E-05)	5.302E-02†[5](9.803E-05)	7.922E-02†[6](1.301E-02)
WFG3.2	2.166E-00†[4]2(6.266E-05)	1.598E-02†[3](5.561E-06)	4.259E-02†[6](7.320E-03)
WFG4.2	1.658E-02†[2](8.878E-07)	1.865E-02†[3](1.539E-06)	1.063E-01†[6](8.903E-03)
WFG5.2	6.839E-02†[2](2.566E-06)	7.297E-02†[6](6.461E-06)	6.749E-02†[4](3.991E-04)
WFG6.2	4.703E-02†[3](1.933E-05)	5.912E-02†[5](3.683E-05)	8.567E-02†[6](1.365E-02)
WFG7.2	1.862E-02†[5](1.237E-06)	1.314E-02†[3](3.478E-07)	2.417E-02†[6](1.276E-03)
WFG8.2	5.574E-02†[5](3.966E-04)	5.583E-02†[6](1.329E-04)	**2.417E-02§[1](1.276E-03)**
WFG9.2	3.671E-02†[4](7.398E-04)	6.546E-02†[5](1.545E-04)	8.308E-02†[6](1.802E-02)
†/§/≈*	23/6/0	26/3/0	26/2/1

(*Continued*)

Table 3.2   (*Continued*)

Instance	MOEA/IGD-NS	DMOEA-εC	DMaOEA-εC
ZDT1	3.824E-03†[4](1.609E-07)	3.763E-03†[2](6.690E-05)	**2.618E-03[1](1.971E-06)**
ZDT2	3.866E-03†[3](1.325E-07)	3.800E-03†[2](7.120E-05)	**2.979E-03[1](1.883E-06)**
ZDT3	2.738E-02†[4](1.365E-03)	5.141E-03†[2](1.696E-04)	**3.435E-03[1](1.582E-04)**
ZDT4	4.387E-03†[4](2.829E-06)	4.060E-03†[3](2.799E-05)	3.901E-03[2](3.334E-05)
ZDT6	3.093E-03†[3](1.065E-06)	3.000E-03†[2](8.050E-05)	**2.200E-03[1](1.718E-06)**
UF1(LZ2)	9.988E-02†[4](5.214E-05)	4.407E-03†[2](9.870E-05)	**2.101E-03[1](4.767E-06)**
UF2(LZ5)	3.483E-02§[6](5.291E-05)	6.503E-03†[2](5.229E-04)	**5.205E-03[1](2.860E-05)**
UF3(LZ8)	1.939E-01†[5](1.873E-03)	**7.209E-03§[1](1.355E-02)**	1.726E-02[2](1.598E-03)
UF4	4.186E-02§[3](5.734E-07)	6.017E-02≈[6](5.017E-03)	6.011E-02[5](5.571E-03)
UF5	2.028E-01§[2](3.908E-04)	3.845E-01†[6](1.251E-01)	2.411E-01[3](1.728E-02)
UF6	**1.336E-01§[1](3.944E-03)**	2.662E-01≈[4](2.012E-01)	2.673E-01[5](1.422E-01)
UF7	3.807E-02†[4](4.976E-04)	4.079E-03†[2](1.269E-03)	**2.512E-03[1](1.901E-03)**
UF8(LZ6)	**2.222E-02§[1](5.901E-03)**	5.284E-02§[3](1.058E-02)	7.081E-02[4](1.195E-03)
UF9	**1.258E-02§[1](2.124E-03)**	4.292E-02≈[4](1.660E-02)	4.220E-02[3](2.081E-03)
UF10	1.005E-01†[3](4.345E-03)	**2.240E-02≈[1](7.361E-02)**	2.251E-02[2](4.769E-03)
LZ1	6.418E-03†[5](2.353E-04)	1.395E-03†[3](1.230E-03)	**8.101E-04[1](2.091E-04)**
LZ3	7.319E-02†[6](4.113E-03)	**2.372E-03§[1](2.794E-03)**	2.454E-03[2](1.470E-03)
LZ4	9.886E-02†[6](8.234E-04)	1.977E-03†[2](3.780E-03)	**1.581E-03[1](3.395E-03)**
LZ7	1.817E-01†[5](3.806E-03)	1.479E-03†[2](2.691E-03)	**1.010E-03[1](1.890E-03)**
LZ9	1.341E-01†[5](1.475E-03)	**3.192E-03§[1](4.437E-02)**	3.579E-03[2](3.410E-02)
WFG1.2	4.246E-01§[3](3.168E-03)	1.063E-00≈[4](1.575E-02)	1.019E-00[5](2.533E-03)
WFG2.2	2.154E-02†[4](2.904E-04)	1.505E-02[2]†(8.403E-04)	**9.381E-03[1](7.562E-04)**
WFG3.2	2.447E-02§[5](5.116E-05)	**1.391E-02§[1](1.199E-03)**	1.519E-02[2](4.105E-05)
WFG4.2	**1.577E-02§[1](6.530E-06)**	4.008E-02†[4](7.401E-03)	4.518E-02[5](3.511E-03)
WFG5.2	6.699E-02†[3](5.068E-06)	6.633E-02†[2](7.970E-04)	**6.578E-02[1](4.332E-04)**
WFG6.2	4.727E-02†[4](1.793E-04)	**2.576E-02§[1](2.087E-04)**	3.189E-02[2](4.098E-04)
WFG7.2	1.359E-02†[4](1.546E-06)	1.082E-02§[2](7.213E-04)	**1.074E-02[1](3.423E-04)**
WFG8.2	5.444E-02†[4](1.195E-04)	3.604E-02§[2](4.116E-03)	3.959E-02[2](1.453E-03)
WFG9.2	**2.137E-02§[1](1.074E-04)**	3.054E-02§[2](1.664E-03)	3.332E-02[2](1.351E-04)
†/§/≈*	22/7/0	16/8/5	

Table 3.3 Statistical Results of 7 Algorithms over 30 Independent Runs on the 29 MOP Instances in Terms of *HV* Metrics.

Instance	NSGA-II	IBEA	MOEA/D
ZDT1	8.708E-01† [5](7.318E-02)	8.710E-01† [4](3.441E-03)	8.686E-01† [6](1.541E-02)
ZDT2	5.371E-01† [5](2.542E-03)	5.366E-01† [6](2.425E-02)	5.421E-01† [3](1.091E-02)
ZDT3	9.879E-01† [5](2.258E-03)	9.014E-01† [6](1.982E-03)	1.003E-00† [2](1.124E-02)
ZDT4	8.696E-01† [4](5.397E-03)	8.226E-01† [6](4.392E-03)	8.896E-01† [2](1.392E-02)
ZDT6	4.326E-01† [4](2.246E-02)	4.314E-01† [6](1.378E-03)	5.042E-01† [3](1.635E-02)
UF1(LZ2)	7.417E-01† [4](3.227E-02)	7.343E-01† [5](6.163E-03)	8.599E-01† [3](9.680E-03)
UF2(LZ5)	8.418E-01† [4](2.269E-04)	8.363E-01† [5](6.460E-03)	8.554E-01† [3](9.815E-03)
UF3(LZ8)	5.997E-01† [4](1.889E-03)	5.945E-01† [5](8.878E-02)	8.428E-01† [3](2.811E-02)
UF4	**4.776E-01§ [1](5.817E-02)**	4.739E-01§ [3](2.024E-03)	4.234E-01† [6](1.298E-02)
UF5	3.579E-01† [2](3.778E-02)	3.115E-01† [4](4.407E-03)	2.206E-01† [5](9.592E-02)
UF6	**4.331E-01§ [1](2.426E-01)**	4.216E-01§ [3](2.137E-02)	3.376E-01† [5](1.055E-01)
UF7	6.374E-01† [6](3.356E-02)	6.465E-01† [4](9.606E-03)	6.939E-01≈ [2](9.662E-03)
UF8(LZ6)	7.563E-01§ [2](8.482E-02)	3.310E-01† [6](0.000E-03)	6.558E-01† [5](2.293E-02)
UF9	1.095E-00§ [2](1.196E-03)	1.009E-00§ [3](5.210E-03)	8.901E-01† [6](7.054E-02)
UF10	5.435E-01† [4](1.268E-02)	3.281E-01† [5](5.030E-03)	1.446E-01† [6](4.563E-02)
LZ1	8.697E-01† [4](4.252E-03)	8.673E-01† [5](4.831E-02)	8.762E-01† [3](7.136E-03)
LZ3	7.908E-01† [5](2.987E-03)	8.024E-01† [4](3.530E-03)	8.664E-01† [3](9.613E-03)
LZ4	7.863E-01† [5](1.349E-02)	8.017E-01† [4](7.082E-03)	8.647E-01† [3](8.462E-03)
LZ7	5.964E-01† [4](3.312E-02)	5.921E-01† [5](3.147E-02)	7.990E-01† [3](8.856E-02)
LZ9	3.824E-01† [4](8.762E-03)	3.073E-01† [6](3.968E-02)	5.213E-01† [3](1.818E-02)
WFG1.2	5.961E-00† [4](4.007E-02)	6.351E-00† [3](1.505E-02)	1.787E-00† [6](6.072E-02)
WFG2.2	5.956E-00† [3](1.168E-03)	5.937E-00≈ [4](5.037E-03)	5.843E-00† [6](7.759E-02)
WFG3.2	5.559E-00† [3](2.453E-02)	**5.594E-00≈ [1](3.364E-02)**	5.430E-00† [6](7.170E-02)
WFG4.2	3.314E-00§ [3](2.756E-03)	**3.336E-00§ [1](5.000E-03)**	2.858E-00† [6](7.019E-02)
WFG5.2	2.996E-00† [3](7.052E-02)	2.992E-00† [4](1.920E-02)	2.981E-00† [6](5.306E-02)
WFG6.2	3.137E-00† [4](5.934E-03)	3.138E-00† [3](4.318E-02)	2.929E-00† [6](1.117E-01)
WFG7.2	3.331E-00† [5](8.840E-03)	3.348E-00† [3](6.144E-02)	3.285E-00† [6](6.356E-02)
WFG8.2	3.071E-00† [5](1.035E-03)	3.089E-00† [3](1.961E-03)	2.692E-00† [6](7.609E-02)
WFG9.2	3.131E-00≈ [3](2.239E-02)	2.995E-00† [5](4.781E-02)	2.911E-00† [6](1.347E-01)
†/§/≈	23/5/1	24/4/1	28/1/0

(*Continued*)

Table 3.3   (*Continued*)

Instance	MOEA/IGD-NS	DMOEA-εC	DMaOEA-εC
ZDT1	8.715E-01†[3](2.155E-03)	8.727E-01†[2](1.541E-02)	**8.735E-01[1](1.073E-03)**
ZDT2	5.373E-01†[4](9.988E-02)	5.434E-01†[2](1.405E-02)	**5.501E-01[1](1.01EE-02)**
ZDT3	9.945E-01†[4](1.990E-03)	1.003E-00†[3](1.304E-02)	**1.0243E-00[1](1.268E-02)**
ZDT4	8.693E-01†[3](2.620E-02)	8.668E-01†[5](4.691E-02)	**8.701E-01[1](3.992E-02)**
ZDT6	4.325E-01†[5](4.980E-03)	5.086E-01†[2](1.469E-02)	**5.346E-01[1](1.083E-02)**
UF1(LZ2)	7.251E-01†[6](1.721E-03)	8.741E-01≈[2](1.434E-02)	**8.742E-01[1](1.882E-02)**
UF2(LZ5)	8.209E-01†[6](8.340E-02)	8.700E-01†[2](1.382E-02)	**8.719e-01[1](2.430E-03)**
UF3(LZ8)	5.549E-01†[6](3.324E-02)	**8.687E-01§[1](1.633E-02)**	8.538E-01[2](1.521E-02)
UF4	4.761E-01§[2](8.706E-02)	4.436E-01≈[4](1.965E-02)	4.342E-01[5](1.063E-02)
UF5	3.315E-01†[3](1.454E-03)	1.256E-01†[6](4.731E-02)	**3.882E-01[1](6.240E-02)**
UF6	4.274E-01§[2](6.037E-02)	3.1704E-01†[6](1.061E-01)	3.882E-01[4](1.804E-02)
UF7	6.430E-01†[5](1.179E-02)	**7.054E-01§[1](1.701E-02)**	6.912E-01[3](1.109E-02)
UF8(LZ6)	**7.742E-01§[1](9.076E-02)**	7.032E-01≈[4](2.423E-02)	7.078E-01[3](1.808E-02)
UF9	**1.107E-00≈[1](8.042E-02)**	1.019E-00≈[4](5.058E-02)	1.011E-00[5](4.112E-02)
UF10	5.512E-01†[3](7.096E-02)	6.088E-01†[2](9.056E-02)	**6.110E-01[1](2.584E-02)**
LZ1	8.648E-01†[6](1.060E-02)	8.797E-01†[2](4.100E-03)	**8.875E-01[1](2.899E-03)**
LZ3	7.688E-01†[6](6.231E-03)	**8.760E-01≈[1](8.710E-03)**	8.710E-01[2](1.125E-02)
LZ4	7.631E-01†[6](8.279E-03)	8.794E-01†[2](7.621E-03)	**8.803E-01[1](6.597E-03)**
LZ7	5.645E-01†[6](4.538E-02)	8.806E-01≈[1](1.350E-02)	8.749E-01[2](2.210E-02)
LZ9	3.656E-01†[5](1.693E-02)	5.322E-01≈[2](2.514E-02)	**5.330E-01[1](2.307E-02)**
WFG1_2	5.294E-00†[5](4.326E-02)	**8.616E-00≈[1](1.512E-02)**	8.591E-00[2](1.191E-02)
WFG2_2	5.926E-00†[5](1.194E-03)	6.076E-00≈[2](1.311E-02)	**6.015E-00[1](2.145E-02)**
WFG3_2	5.531E-00†[4](1.587E-02)	5.472E-00†[2](1.541E-02)	5.584E-00[2](3.224E-02)
WFG4_2	3.311E-00§[2](6.240E-03)	3.105E-00§[4](1.645E-02)	3.083E-00[5](1.008E-02)
WFG5_2	2.982E-00†[5](9.190E-02)	3.025E-00†[2](1.410E-02)	**3.193E-00[1](3.338E-02)**
WFG6_2	3.127E-00†[5](3.757E-02)	**3.303E-00≈[1](2.479E-02)**	3.292E-00[2](1.569E-03)
WFG7_2	3.334E-00†[4](1.003E-02)	3.351E-00†[2](1.622E-02)	**3.435E-00[1](6.981E-03)**
WFG8_2	3.074E-00†[4](3.100E-03)	**3.340E-00§[1](1.605E-02)**	3.162E-00[2](1.113E-02)
WFG9_2	3.191E-00§[1](6.858E-02)	3.138E-00§[2](2.067E-02)	3.088E-00[4](2.242E-02)
†/§/≈	23/5/1	14/5/10	

competitive performance on all LZ test problems. DMaOEA-εC outperforms comparison algorithms on all bi-objective WFG problems except WFG1 and WFG4, on which IBEA and MOEA/IGD-NS perform better. The reason for the unsatisfactory performance of DMaOEA-εC on WFG4 and WFG9 problems might be that DMaOEA-εC is not powerful at tackling MOPs with degenerate or deceptive properties. To sum up, Table 3.3 shows that DMaOEA-εC achieves significantly better *HV* values than NSGA-II, IBEA, MOEA/D, MOEA/IGD-NS and DMOEA-εC in 23, 24, 28, 23 and 13 out of the 29 test instances, respectively. In conclusion, DMaOEA-εC shows better performance on the majority of MOPs.

3.3.2 *Comparisons on MaOPs*

This part of the experiment is designed to study the effectiveness of DMaOEA-εC on MaOPs. The means and standard deviations of *IGD* and *HV* metric values over 30 independent runs of each algorithm on 16 test instances are shown in Tables 3.4 and 3.5, respectively.

As can be seen from Table 3.4, in terms of *IGD* metric values, DMaOEA-εC shows a significant advantage over competitors on all DTLZ1 instances. DMaOEA-εC performs the best on the DTLZ2 with three objectives. Besides, Two_Arch2 shows superiority over DMaOEA-εC on DTLZ2 with 5 and 8 objectives; MOEADD performs better than the DMaOEAεC on DTLZ2 with 10 objectives. DMaOEA-εC shows a clear advantage over others on DTLZ3 with 3 and 10 objectives. In spite of that, DMaOEA-εC is not the top among all the algorithms on the rest of the DTLZ3 test problems, the rank values of DMaOEA-εC on the remaining DTLZ3 problems are rightly after Two_Arch2. As to the DTLZ4 test instances, DMaOEA-εC performs significantly better than its competitors on all DTLZ4 test problems except for the DTLZ4 with four objectives on which NSGA-III shows the best performance. To summarize, as can be seen in Table 3.4, DMaOEAεC achieves significantly better *IGD* values than HypE, NSGA-III, MOEADD and Two_Arch2 in 16, 15, 10, 12 and 13 out of the 16 test instances, respectively.

Experimental results on DTLZ problems in terms of *HV* metrics show that DMaOEAεC performs significantly better than its competitors on all DTLZ1 and DTLZ4 instances. As to the DTLZ2 test problems, DMaOEAεC performs best on DTLZ2 with three objectives, but it shows worse performances than its MOEADD on the remaining DTLZ2

Table 3.4   Statistical Results of 7 Algorithms over 30 Independent Runs on the 16 MaOP DTLZ Instances in Terms of IGD Metrics.

Instance	HypE	NSGA-III	MOEADD
DTLZ1_3	3.275E-01†[7](7.086E-05)	3.147E-01†[2](1.952E-04)	3.155E-01†[4](7.013E-05)
DTLZ1_5	3.730E-01†[6](2.845E-03)	2.592E-01†[2](7.732E-03)	2.694E-01†[5](1.875E-03)
DTLZ1_8	4.341E-01†[6](3.231E-04)	2.347E-01†[4](1.201E-04)	2.298E-01†[2](8.953E-05)
DTLZ1_10	4.111E-01†[6](8.363E-03)	2.345E-01†[4](1.287E-03)	2.331E-01†[3](7.036E-04)
DTLZ2_3	1.300E-01†[6](4.143E-03)	5.494E-02†[4](3.072E-04)	5.456E-02≈[2](1.611E-04)
DTLZ2_5	3.480E-01†[5](1.157E-04)	1.797E-01†[4](1.352E-03)	1.697E-01§[2](9.451E-04)
DTLZ2_8	6.208E-01†[6](3.242E-03)	4.634E-01†[5](1.343E-02)	3.597E-01§[2](6.971E-04)
DTLZ2_10	7.756E-01†[6](2.285E-02)	4.730E-01†[5](1.207E-02)	**3.919E-01§[1](3.732E-03)**
DTLZ3_3	3.236E-01†[6](5.379E-03)	5.488E-02†[2](1.771E-04)	5.488E-02†[2](2.108E-04)
DTLZ3_5	6.786E-01†[6](6.811E-03)	1.697E-01†[3](1.552E-03)	1.697E-01†[5](3.008E-03)
DTLZ3_8	9.186E-01†[6](2.257E-03)	4.335E-01†[5](1.054E-02)	**3.593E-01§[1](5.104E-03)**
DTLZ3_10	8.709E-01†[6](2.779E-03)	5.898E-01†[5](1.317E-02)	3.918E-02†[3](2.987E-03)
DTLZ4_3	6.383E-01†[6](8.063E-03)	2.500E-01†[4](7.136E-02)	5.487E-02†[2](4.040E-04)
DTLZ4_5	4.702E-01†[6](3.347E-03)	1.700E-01†[3](1.306E-03)	1.697E-01≈[2](4.280E-04)
DTLZ4_8	6.389E-01†[6](9.169E-03)	3.600E-01≈[2](1.753E-03)	3.811E-01†[5](2.377E-03)
DTLZ4_10	9.208E-01†[6](5.657E-03)	7.464E-01†[5](1.145E-03)	5.796E-01†[2](5.882E-03)
†/§/≈	16/0/0	15/0/1	10/4/2

Table 3.4   (*Continued*)

Instance	Two_Arch2	θ-DEA	DMaOEA-εC
DTLZ1.3	3.168E-01[†][6](9.132E-04)	3.154E-01[†][3](4.949E-05)	**5.061E-02**[1]**(2.712E-05)**
DTLZ1.5	2.666E-01[†][4](4.334E-04)	2.594E-01[†][3](1.106E-03)	**1.099E-01**[1]**(4.148E-03)**
DTLZ1.8	2.456E-01[†][5](6.341E-03)	2.300E-01[†][3](5.122E-03)	**1.917E-01**[1]**(2.358E-03)**
DTLZ1.10	2.584E-01[†][5](6.614E-04)	2.330E-01[†][2](1.326E-03)	**2.317E-01**[1]**(4.322E-04)**
DTLZ2.3	6.108E-02[†][5](1.443E-03)	5.489E-02[†][3](5.016E-04)	**5.443E-02**[1]**(7.133E-04)**
DTLZ2.5	**1.696E-01**[§][1]**(4.728E-04)**	3.604E-01[†][6](3.611E-04)	1.728E-01[3](1.830E-04)
DTLZ2.8	**3.368E-01**[§][1]**(7.322E-03)**	3.604E-01[≈][3](9.611E-03)	3.786E-01[4](3.809E-04)
DTLZ2.10	4.070E-01[†][4](4.379E-03)	3.924E-01[≈][2](1.454E-03)	4.018E-01[3](2.071E-03)
DTLZ3.3	6.067E-02[†][5](2.831E-04)	5.489E-02[†][4](6.066E-04)	**5.120E-02**[1]**(2.429E-04)**
DTLZ3.5	1.672E-01[§][1](1.733E-03)	1.697E-01[†][4](2.892E-03)	1.689E-01[2](2.974E-03)
DTLZ3.8	3.629E-01[§][2](7.663E-03)	4.315E-01[†][4](2.596E-02)	4.304E-01[3](1.919E-03)
DTLZ3.10	4.057E-01[†][4](1.712-03)	3.916E-01[†][2](3.867E-03)	**3.777E-01**[1]**(3.122E-03)**
DTLZ4.3	2.367E-01[†][3](1.570E-01)	2.508E-01[†][5](7.197E-02)	**5.470E-01**[1]**(1.710E-04)**
DTLZ4.5	1.702E-01[†][4](8.253E-03)	**3.596E-01**[†][5]**(2.067E-04)**	**1.677E-01**[1]**(2.226E-03)**
DTLZ4.8	3.371E-01[†][4](8.435E-03)	**3.596E-01**[§][1]**(2.066E-03)**	3.663E-01[3](1.995E-03)
DTLZ4.10	6.097E-01[†][4](7.996E-03)	5.809E-01[†][3](1.119E-04)	**4.851E-01**[1]**(4.770E-03)**
†/§/≈	12/4/0	13/1/2	

Table 3.5   Statistical Results of 7 Algorithms over 30 Independent Runs on the 16 MaOP DTLZ Instances in Terms of *HV* Metrics.

Instance	HypE	NSGA-III	MOEADD
DTLZ1_3	1.272E-00†[5](3.504E-03)	1.204E-00†[6](1.248E-03)	1.305E-00†[2](3.716E-03)
DTLZ1_5	1.535E-00†[6](5.786E-02)	1.609E-00†[3](1.616E-03)	1.605E-00†[5](1.033E-04)
DTLZ1_8	1.954E-00†[6](1.503E-02)	2.143E-00†[2](1.418E-03)	2.142E-00†[4](6.885E-04)
DTLZ1_10	2.526E-00†[6](1.815E-02)	2.591E-00†[3](1.343E-04)	2.591E-00†[4](2.015E-04)
DTLZ2_3	7.11E-01†[6](4.580E-02)	7.442E-01†[4](1.027E-03)	7.444E-01≈[2](1.415E-04)
DTLZ2_5	1.159E-00†[6](1.850E-02)	1.307E-00§[3](1.789E-03)	**1.407E-00§[1](3.312E-02)**
DTLZ2_8	1.696E-00†[5](8.806E-03)	1.850E-00†[4](2.116E-02)	**1.980E-00§[1](2.740E-03)**
DTLZ2_10	1.940E-00†[6](2.380E-02)	2.426E-00†[4](1.461E-02)	**2.515E-00§[1](1.132E-03)**
DTLZ3_3	3.160E-01†[6](2.361-02)	6.269E-01†[3](2.593E-02)	7.448E-01≈[2](2.197E-03)
DTLZ3_5	5.117E-01†[6](8.282E-02)	1.280E-00†[4](1.858E-03)	1.308E-00≈[2](7.166E-03)
DTLZ3_8	6.662E-01†[6](5.303E-02)	1.831E-00†[3](4.394E-02)	**1.980E-00§[1](5.482E-03)**
DTLZ3_10	1.239E-00†[6](9.571E-01)	2.206E-00†[5](4.27E-02)	2.515E-00≈[2](2.731E-03)
DTLZ4_3	3.160E-01†[6](2.364E-02)	6.269E-01†[3](2.593E-02)	7.448E-01†[2](2.197E-03)
DTLZ4_5	1.094E-00†[6](5.285E-02)	1.308E-00†[4](5.489E-03)	1.309E-00†[3](5.228E-03)
DTLZ4_8	1.787E-00†[5](5.353E-02)	1.975E-00†[4](2.502E-03)	1.963E-00≈[3](1.475E-02)
DTLZ4_10	2.039E-00†[6](2.214E-02)	2.515E-00†[4](6.995E-03)	2.517E-00†[2](1.998E-03)
†/§/≈	16/0/0	15/1/0	7/4/5

Table 3.5 (*Continued*)

Instance	Two_Arch2	θ-DEA	DMaOEA-εC
DTLZ1_3	1.302E-00†[4](2.319E-03)	1.304E-00†[3](2.696-03)	**1.524E-00[1](1.736E-03)**
DTLZ1_5	1.608E-00†[4](2.060E-03)	1.609E-00†[2](2.504E-03)	**1.641E-00[1](3.631E-03)**
DTLZ1_8	2.142E-00†[3](3.763E-03)	2.141E-00†[5](4.039E-03)	**2.156E-00[1](3.911E-04)**
DTLZ1_10	2.591E-00†[5](3.197E-03)	2.593E-00†[2](1.343E-04)	**2.608E-00[1](1.499E-04)**
DTLZ2_3	7.409E-01†[5](2.499E-02)	7.444E-01†[3](1.534E-03)	**7.458E-01[1](1.113E-04)**
DTLZ2_5	1.236E-00†[5](5.997E-03)	1.307E-00§[2](1.760E-03)	1.243E-00[4](2.116E-03)
DTLZ2_8	1.608E-00†[6](1.316E-03)	1.980E-00§[2](4.530E-03)	1.877E-00[3](2.642E-03)
DTLZ2_10	2.052E-00†[5](1.937E-03)	2.515E-00≈[2](1.217E-03)	2.485E-00[3](1.469E-03)
DTLZ3_3	6.177E-01†[5](7.711E-02)	6.19EF-01†[4](2.946E-02)	**7.450E-01[1](3.336E-03)**
DTLZ3_5	1.270E-00†[5](2.448E-03)	**1.308E-00≈[1](4.020E-03)**	1.305E-00[3](2.702E-03)
DTLZ3_8	1.768E-00†[5](2.232E-03)	1.812E-00†[4](1.411E-01)	1.961E-00[2](1.547E-03)
DTLZ3_10	2.225E-00†[4](1.036E-03)	2.515E-00≈[3](7.044E-03)	**2.519E-00[1](4.732E-03)**
DTLZ4_3	6.177E-01†[5](7.711E-02)	6.194E-01†[4](2.946E-02)	**7.487E-01[1](3.363E-03)**
DTLZ4_5	1.227E-00†[5](2.915E-02)	**1.980E-00§[1](1.397E-03)**	1.378E-00[2](2.779E-03)
DTLZ4_8	1.623E-00†[6](1.186E-03)	**1.980E-00§[1](1.397E-03)**	1.976E-00[2](6.079E-03)
DTLZ4_10	2.056E-00†[5](9.229E-03)	2.515E-00†[3](1.846E-03)	**2.526E-00[1](3.053E-03)**
†/§/≈	16/0/0	9/4/3	

problems. DMaOEA-εC outperforms other competitors on DTLZ3 with 3 and 10 objectives. As to the remaining DTLZ3 test problems, the performance of DMaOEAεC is competitive with MOEADD. For *HV* metric values, DMaOEAεC performs significantly better than these competitors in 16, 15, 7, 16 and 9 out of the 16 instances, respectively.

The experimental results on WFG test suites in Table 3.6 show that DMOEA-εC is not the top among all the algorithms on all WFG1 on which HyPE and Two_Arch2 perform better. DMaOEA-εC performs significantly better than competitors on all WFG2 and WFG3 problems. DMaOEA-εC exhibits the best performance on WFG4 with three objectives and performs slightly worse than θ-DEA and NSGA-III on the remaining WFG4 problems. For WFG5, DMaOEA-εC has clear advantages over competitors on WFG5 with 3 and 5 objectives and performs competively with others on WFG with 8 and 10 objectives on which θ-DEA and NSGA-III perform better. DMaOEA-εC outperforms or performs competitively against comparisons on the majority of WFG6 instances. To be specific, HyPE and θ-DEA show slightly better performance than DMaOEA-εC on WFG6 with 3/5 and 8/10 objectives, respectively. As to WFG7, DMaOEA-εC exhibits the best performance on all WFG7 instances except WFG7 with five objectives, on which θ-DEA and NSGA-III show better performance. DMaOEA-εC has an obvious advantage on WFG8 with 8 and 10 objectives and only shows a slightly worse performance after NSGA-III. The performance of DMaOEA-εC on WFG9 with 3 and 5 objectives is significantly better over others. However, θ-DEA and NSGA-III show the best performance on WFG with 8 and 10 objectives and the performance of DMaOEA-εC is rightly after them. To sum up, DMaOEA-εC achieves significantly better *IGD* values than HyPE, NSGA-III, MOEADD and Two_Arch2 in 25, 24, 36, 30 and 20 out of the 36 test instances, respectively.

In conclusion, DMaOEA-εC exhibits the best performance on all DTLZ1 and the majority of DTLZ4 instances. As to DTLZ2 and DTLZ3, DMaOEA-εC shows medium performance. What's more, Two_Arch2 and MOEADD show better performances in terms of *IGD* values. MOEADD and θ-DEA exhibit better performance in terms of *HV* metrics. For WFG instances, DMaOEA-εC exhibits various performance on different instances and has advantages over comparison algorithms.

Table 3.6 Statistical Results of 7 Algorithms over 30 Independent Runs on the 36 MaOP WFG Instances in Terms of *HV* Metrics.

Instance	HypE	NSGA-III	MOEADD
WFG1_3	4.064E+01† [5](1.483E-01)	4.647E+01† [4](2.486E-00)	3.269E+01† [6](1.956E+01)
WFG1_5	**5.946E+03§ [1](5.530E+03)**	4.891E+03† [6](2.036E+04)	4.937E+03† [5](8.291E+03)
WFG1_8	2.194E+07§ [1](2.304E+10)	1.964E+07† [5](9.435E+10)	1.717E+07† [6](3.567E+12)
WFG1_10	9.599E+09† [2](5.817E+13)	9.553E+09≈ [3](2.248E+16)	9.412E+07† [6](9.366E+16)
WFG2_3	**5.967E+01† [1](3.828E-01)**	5.880E+01† [5](6.745E-02)	5.852E+01† [6](1.524E-01)
WFG2_5	6.164E+03≈ [2](1.002E+02)	6.137E+03† [4](9.017E+01)	6.029E+03† [6](3.927E+02)
WFG2_8	2.208E+07† [2](8.002E+08)	2.190E+07† [5](5.999E+09)	2.113E+07† [6](1.055E+10)
WFG2_10	9.629E+09† [2](8.949E+13)	9.460E+09† [4](1.327E+16)	9.052E+09† [5](4.070E+15)
WFG3_3	4.079E+01† [2](6.127E-01)	3.945E+01† [6](2.280E-01)	3.710E+01† [6](2.280E-00)
WFG3_5	4.122E+03† [2](1.183E+03)	3.923E+03† [5](5.524E+03)	3.739E+03† [6](2.130E+03)
WFG3_8	1.448E+07† [2](9.273E+10)	1.356E+07† [4](6.263E+11)	1.169E+07† [5](2.897E+11)
WFG3_10	6.311E+09† [2](8.580E+15)	5.542E+09† [4](1.175E+17)	3.863E+09† [6](4.056E+15)
WFG4_3	3.554E+01≈ [2](7.344E-01)	3.485E+01† [5](1.121E-02)	3.438E+01† [6](8.718E-03)
WFG4_5	**4.907E+03§ [1](2.457E+01)**	4.921E+03≈ [3](8.255E+01)	4.807E+03† [5](9.289E+01)
WFG4_8	1.672E+07† [6](5.688E+09)	2.010E+07§ [2](5.996E+09)	1.857E+07† [4](4.089E+10)
WFG4_10	6.763E+09† [6](6.616E+10)	8.718E+09† [3](5.741E+17)	7.122E+09† [5](1.771E+16)
WFG5_3	3.301E+01† [4](9.388E-02)	3.306E+01≈ [2](5.319E-04)	3.249E+01† [5](1.589E-03)
WFG5_5	4.698E+03† [4](5.847E+02)	4.699E+03† [3](2.88E+01)	4.560E+03† [5](4.024E-01)
WFG5_8	1.751E+07† [4](7.780E+07)	1.905E+07§ [2](1.420E+08)	1.734E+07† [5](1.317E+10)
WFG5_10	7.322E+09† [5](9.047E+10)	**8.435E+09§ [1](4.373E+12)**	6.273E+09† [6](1.758E+16)
WFG6_3	3.285E+01≈ [2](4.940E-02)	3.227E+01† [5](1.132E-01)	3.191E+01† [6](8.925E-01)
WFG6_5	**4.682E+03† [1](2.494E+03)**	4.659E+03† [4](1.795E+03)	4.514E+03† [5](7.907E+02)
WFG6_8	1.873E+07≈ [3](2.593E+09)	1.868E+07† [4](6.172E+10)	1.714E+07† [5](1.331E+11)
WFG6_10	6.754E+09† [4](6.138E+12)	6.105E+09† [6](3.195E+18)	6.118E+09† [5](1.252E+17)
WFG7_3	3.546E+01† [2](3.908E-03)	3.520E+01† [5](8.655E-03)	3.454E+01† [6](7.085E-03)
WFG7_5	4.950E+03§ [3](4.831E+02)	4.963E+03§ [2](1.023E+02)	4.807E+03† [5](7.192E+02)
WFG7_8	1.856E+07† [5](2.178E+10)	2.023E+07† [3](1.163E+09)	1.918E+07† [4](3.533E+10)
WFG7_10	7.920E+09† [3](1.293E+15)	8.613E+09≈ [2](5.352E+17)	7.204E+09† [5](2.235E+16)
WFG8_3	3.141E+01† [4](1.695E-02)	**3.177E+01§ [1](3.027E-02)**	3.117E+01† [5](7.219E-03)
WFG8_5	4.467E+03† [4](7.780E+01)	**4.473E+03§ [1](6.751E+01)**	4.372E+03† [5](3.076E+02)
WFG8_8	1.736E+07† [3](1.544E+10)	1.693E+07† [3](6.797E+11)	1.667E+07† [6](4.761E+11)
WFG8_10	7.840E+09† [3](4.209E+13)	6.645E+09† [5](1.244E+17)	7.445E+09† [4](7.823E+15)
WFG9_3	**3.483E+01≈ [1](1.861E-02)**	3.120E+01† [6](3.642E-00)	3.234E+01† [4](4.708E-01)
WFG9_5	4.909E+03† [2](1.100E-03)	4.480E+03† [5](4.761E+04)	4.408E+03† [6](2.166E+03)
WFG9_8	1.574E+07† [5](3.874E+10)	1.766E+07§ [2](1.453E+12)	1.544E+07† [6](1.054E+12)
WFG9_10	6.571E+09† [5](1.77E+16)	7.632E+09§ [2](4.801E+17)	5.531E+09† [6](1.253E+17)
†/§/≈	25/4/7	24/8/4	36/0/0

(*Continued*)

Table 3.6 (*Continued*)

Instance	Two_Arch2	θ-DEA	DMaOEAεC
WFG1.3	4.727E+01≈[2](2.894E-00)	**4.739E+01§[1](6.989E-01)**	4.716E+01[3](7.248E-00)
WFG1.5	5.727E+03§[2](1.691E+04)	5.159E+03†[4](3.365E+04)	5.592E+03[3](2.226E+03)
WFG1.8	2.182E+07†[2](4.562E+10)	2.138E+07≈[3](8.484E+11)	2.113E+07[4](7.794E+10)
WFG1.10	**9.612E+09§[1](1.028E+13)**	9.424E+09§[5](2.710E+15)	9.520E+09[4](1.439E+15)
WFG2.3	5.908E+01†[4](4.561E-02)	5.925E+01†[3](2.163E-02)	5.945E+01[2](7.985E-03)
WFG2.5	6.146E+03†[3](3.552E+01)	6.135E+03†[5](5.552E+01)	6.168E+03[1](4.303E+02)
WFG2.8	2.200E+07†[3](9.479E+08)	2.194E+07†[4](2.079E+09)	**2.212E+07[1](4.039E+08)**
WFG2.10	9.623E+09†[3](1.842E+12)	8.720E+09†[6](8.760E+17)	**9.637E+09[1](4.569E-13)**
WFG3.3	3.992E+01†[3](7.957E-02)	3.946E+01†[4](7.626E-02)	**4.908E+01[1](1.655E-02)**
WFG3.5	4.056E+03†[3](3.500E+02)	3.931E+03†[4](8.590E+02)	**4.285E+03[1](7.147E-02)**
WFG3.8	1.426E+07†[3](1.621E+10)	1.164E+07†[6](5.566E+11)	**1.586E+07[1](6.539E-10)**
WFG3.10	6.188E+09†[3](2.323E+15)	5.138E+09†[5](9.052E+17)	**6.911E+09[1](5.708E-15)**
WFG4.3	3.490E+01†[4](1.050E-02)	3.497E+01†[3](2.185E-02)	**3.582E+01[1](1.263E-01)**
WFG4.5	4.638E+03§[6](1.546E+03)	4.923E+03§[2](5.621E+01)	4.829E+03[4](7.743E+01)
WFG4.8	1.767E+07§[5](1.557E+10)	**2.013E+07§[1](4.363E-09)**	1.880E+07[3](3.467E+09)
WFG4.10	7.998E+09†[4](6.845E+15)	**9.060E+09§[1](8.175E-12)**	8.722E+09[2](2.684E+10)
WFG5.3	3.247E+01†[6](6.271E-02)	3.306E+01≈[3](7.992E-04)	**3.307E+01[1](6.262E-05)**
WFG5.5	4.410E+03†[6](8.780E+02)	4.702E+03†[2](7.315E-00)	**4.760E+03[1](7.248E-00)**
WFG5.8	1.639E+07†[6](1.478E+10)	**1.906E+07§[1](1.602E-08)**	1.778E+07[3](5.343E+08)
WFG5.10	7.373E+09†[4](1.098E+15)	8.434E+09§[2](2.904E+12)	7.658E+09[3](1.114E+12)
WFG6.3	3.266E+01†[3](3.637E-01)	3.262E+01†[4](2.136E-02)	**3.288E+01[1](6.061E-02)**
WFG6.5	4.356E+03†[6](4.104E+02)	4.664E+03†[3](2.689E+02)	4.674E+03[2](4.114E+02)
WFG6.8	1.651E+07†[6](2.563E+10)	1.876E+07≈[2](5.256E+10)	**1.877E+07[1](5.343E-10)**
WFG6.10	7.283E+09§[3](2.529E+17)	**8.103E+09§[1](1.660E-16)**	7.45E+09[2](3.559E+15)
WFG7.3	3.527E+01†[4](1.878E-03)	3.531E+01†[3](5.331E-04)	**3.617E+01[1](4.383E-03)**
WFG7.5	4.772E+03†[6](2.069E+02)	**4.980E+03§[1](3.393E-01)**	4.876E+03[4](5.368E+03)
WFG7.8	1.801E+07†[6](2.589E+10)	2.029E+07≈[2](5.562E+08)	**2.046E+07[1](2.409E+09)**
WFG7.10	7.680E+09†[4](1.000E+17)	9.054E+07†[6](2.221E+10)	**8.611E+09[1](3.669E-16)**
WFG8.3	3.105E+01†[6](2.749E-02)	3.170E+01≈[2](1.66E-02)	3.163E+01[3](1.955E-03)
WFG8.5	4.143E+03†[6](2.220E+01)	4.447E+03†[3](1.513E-02)	4.481E+01[2](7.083E+01)
WFG8.8	1.421E+07†[6](6.480E+10)	**1.767E+07≈[1](1.089E+10)**	1.740E+07[2](1.052E+10)
WFG8.10	5.854E+09§[6](1.085E+16)	7.848E+09†[2](4.033E+14)	**7.924E+09[1](3.278E-14)**
WFG9.3	3.277E+01§[5](3.209E-00)	3.320E+01†[3](1.873E-01)	3.477E+01[2](1.678E-01)
WFG9.5	4.483E+03†[4](7.311E+02)	4.689E+03†[3](1.254E+02)	5.011E+03[1](7.804E+03)
WFG9.8	1.583E+07†[4](1.023E+11)	1.821E+07§[1](1.110E+12)	1.602E+07[3](1.692E+10)
WFG9.10	6.933E+09§[4](1.655E+16)	**8.463E+09§[1](1.353E-16)**	7.591E+09[3](2.491E+15)
†/§/≈	33/2/1	20/10/6	

3.4   Detailed Analysis on Behavior of DMaOEA-εC

In this section, the algorithmic behavior of DMaOEA-εC is deeply analyzed. First, the superiority of the two-side update rule is investigated. Then, effects of the boundary points maintenance mechanism and the distance-based global replacement strategy are further investigated.

3.4.1   *Superiority of the Two-side Update Rule*

As mentioned above, the proposed two-side update rule that maintains both feasible and infeasible solutions for each subproblem optimizes each subproblem from both feasible and infeasible sides. It does not increase any additional computational cost, but only needs more memory space. This update procedure can maintain more information on the population, make the update more effective, and thus enhance the convergence. Then, with the aim of showing the superiority of the proposed two-side update rule over the original feasibility rule, we develop a variant of DMOEA-εC, namely DMOEA-εC_TR, for the comparison with the original DMOEA-εC. In DMOEA-εC_TR, the feasibility rule is replaced with the proposed two-side update rule.

Since DMOEA-εC_TR is only applicable for MOPs, the ZDT1, WFG2_2, UF1, DTLZ1_3 and WFG3_3 are selected as test instances. With the same parameter settings as Section 3.2, DMOEA-εC_TR is experimentally compared with DMOEA-εC. The experimental results, in terms of the means and standard deviations of *IGD* and *HV* metric values within 30 independent runs obtained by each algorithm for the selected test instances are all shown in Table 3.7.

Table 3.7 shows that in terms of *IGD* and *HV* metrics, DMOEA-εC_TR is significantly better than DMOEA-εC on all selected instances. The superiority of the proposed two-side update rule is confirmed experimentally.

3.4.2   *Effects of the Boundary Maintenance Mechanism and the Distance-based Global Replacement Strategy*

With the aim of showing the effect of the boundary points maintenance mechanism, a variant of DMaOEA-εC without the boundary points maintenance mechanism, denoted as DMaOEA-εC_NoB, is devel-

Table 3.7   Statistical Results (Mean (Std. Dev.)) of DMOEA-εC and DMOEA-εC_TR over 30 Independent Runs on the 5 Selected MOPs in Terms of *IGD* and *HV* Metrics.

Instance	DMOEA-εC	DMOEA-εC_TR
	IGD	
ZDT1	3.763E-03(6.690E-05)†	**3.419E-03(2.224E-06)**
UF1	4.407E-03(2.9870E-05)†	**3.676E-03(3.129E-06)**
WFG2_2	1.505E-02(8.403E-04)†	**1.081E-02(2.287E-04)**
DTLZ1_3	**4.976E-02(9.966E-04)**§	5.061E-02(2.1712E-05)
WFG3_3	-	-
	HV	
ZDT1	8.727E-01(1.541E-02)†	**8.732E-01(4.110E-03)**
UF1	8.742E-01(1.434E-02)$\approx$	**8.741E-01(1.882E-02)**
WFG2_2	6.006E-00(1.311E-02)†	**6.015E-00(2.145E-02)**
DTLZ1_3	1.307E-00(4.458E-03)†	**1.414E-00(3.710E-03)**
WFG3_3	3.071E+01(8.352E-02)†	**4.908E+01(1.655E-02)**

oped. Similarly, we replace the distance-based global replacement strategy in DMaOEA-εC with the original local replacement strategy — and denote it as DMaOEA-εC_LR. Here we still consider ZDT1, WFG2_2, UF1, DTLZ1_3 and WFG3_3 as multi-objective test instances. Besides, DTLZ2_5, DTLZ3_8, DTLZ4_10, WFG5_5, WFG7_8 and WFG8_10 are selected as many-objective test problems. With the same parameter settings as Section 3.2, DMaOEA-εC is experimentally compared with its variants DMaOEA-εC_NoB and DMaOEA-εC_LR. The experimental results, in terms of the means and standard deviations of *IGD* and *HV* metric values within 30 independent runs obtained by each algorithm for the selected test instances are all shown in Table 3.8.

Experimental results as shown in Table 3.8 demonstrate that the proposed DMaOEA-εC shows remarkable advantages over DMaOEA-εC_NoB on all selected instances, especially on selected many-objective instances, in terms of *IGD* and *HV* metrics instances. Similarly, Table 3.8 shows the better performance of DMaOEA-εC over DMaOEA-εC_LR. These experimental results confirm the effectiveness of the boundary maintenance mechanism and the distance-based global replacement strategy.

Table 3.8   Statistical Results (Mean (Std. Dev.)) of DMaOEA-εC, DMaOEA-εC_NoB, and DMaOEA-εC_LR over 30 Independent Runs on the 12 Selected MOPs and MaOPs in Terms of *IGD* and *HV* Metrics.

Instance	DMaOEA-εC	DMaOEA-εC_NoB	DMaOEA-εC_LR
		IGD	
ZDT1	**2.618E-03(1.971E-06)**	2.761E-03(4.563E-06)†	2.722E-03(1.818E-05)†
UF1	**2.101E-03(4.767E-06)**	4.423E-03(1.616E-05)†	4.407E-03(8.050E-06)†
WFG2_2	**9.381E-03(7.562E-04)**	1.306E-01(9.057E-04)†	1.112E-01(9.193E-04)†
DTLZ1_3	**5.061E-02(2.712E-05)**	5.312E-02(2.868E-05)†	5.476E-02(1.629E-05)†
DTLZ2_5	**3.768E-01(3.809E-04)**	3.856E-01(5.924E-04)†	3.901E-01(6.225E-04)†
DTLZ3_8	**4.304E-01(1.919E-03)**	4.994E-01(2.188E-03)†	4.749E-01(1.670E-03)†
DTLZ4_10	**4.851E-01(4.770E-03)**	5.468E-01(9.575E-04)†	4.920E-01(3.683E-03)†
		HV	
ZDT1	**8.735E-01(1.073E-03)**	8.677E-01(1.607E-03)†	8.686E-01(1.541E-03)†
UF1	**8.742E-01(1.882E-03)**	7.204E-01(6.371E-06)†	8.056E-01(1.759E-06)†
WFG2_2	**6.015E-00(2.145E-02)**	5.905E-00(7.075E-03)†	5.843E-00(6.190E-02)†
DTLZ1_3	**1.524E-00(1.736E-03)**	1.350E-00(1.691E-03)†	1.269E-00(3.375E-03)†
DTLZ2_5	**1.243E-00(2.116E-03)**	9.632E-01(1.410E-02)†	1.023E-00(9.754E-03)†
DTLZ3_8	**1.961E-00(1.547E-01)**	1.278E-00(1.480E-01)†	1.498E-00(1.100E-01)†
DTLZ4_10	**2.526E-00(3.053E-03)**	2.151E-00(5.430E-03)†	2.498E-00(4.298E-03)†
WFG3_3	**4.908E+01(1.655E-02)**	3.964E+01(5.351E-02)†	3.888E+01(2.770E-02)†
WFG5_5	**4.760E+03(1.248E+02)**	4.157E+03(6.167E+03)†	4.130E+01(7.548E+03)†
WFG7_8	**2.046E+07(2.409E+09)**	1.970E+07(5.176E+08)†	1.927E+07(6.160E+09)†
WFG8_10	**7.924E+09(3.278E+14)**	6.948E+09(1.624E+14)†	6.957E+09(2.946E+13)†

3.5   Conclusion

Decomposition and the ε-constraint method are two important strategies in the field of multi-objective optimization. DMOEA-εC gives the first attempt to incorporate the ε-constraint method into the decomposition strategy and solve an MOP via optimizing a series of scalar-constrained subproblems collaboratively using the neighbor information. Since DMOEA-εC lacks the ability of dealing with MaOPs, this chapter proposes an improved version of DMOEA-εC, named DMaOEA-εC. Firstly, a two-side update rule is proposed to maintain both feasible and infeasible solutions for each subproblem and enhance the convergence. Then, with the aim of overcoming ineffectiveness induced by the exponential number of upper bound vectors, we present a systematic two-stage approach to generate widely spread upper bound vectors in a high-dimensional space. Besides, in order to remedy the diversity loss of a population in decomposition-based approaches, a distance-based global replacement strategy is put forward to compare a new generated solution with all subproblems in a distance-based order. What's more, a boundary point maintenance mechanism is introduced by taking advantage of corner solutions to reduce the possibility of losing boundary solutions and thus achieve a good spread of solutions over the PF. The above-mentioned mechanisms are incorporated into DMOEA-εC, which results in DMaOEA-εC.

DMaOEA-εC has been compared with five state-of-the-art and classical mult-objective evolutionary algorithms, including NSGA-II, IBEA, MOEA/D, MOEA/IGD-NS and DMOEA-εC on 29 test instances to exhibit its performance on MOPs. Furthermore, DMaOEA-εC has been compared with four state-of-the-art many-objective evolutionary algorithms, including HypE, NSGA-III, MOEADD and Two_Arch2, to exhibit its performance on MaOPs. A systematical experimental study has demonstrated that DMaOEA-εC outperforms or performs competitively against other algorithms on the majority of MOPs and MaOPs with up to 10 objectives.

Chapter 4

An *A Posteriori* Decision-making Framework and Subproblems Co-solving Evolutionary Algorithm for Uncertain Optimization

In one sense, uncertain optimization problems (UOPs) are similar to multi-objective problems (MOPs). That is to say, according to the phase in which decision makers (DMs) provide their preferences on uncertainties in the optimization process, approaches for dealing with UOPs can also be classified into three groups: *a priori* methods, interactive methods and *a posteriori* methods [Miettinen (1999)]. In *a priori* methods, DMs present preferences on uncertainties before the optimization process, and then UOPs can be converted into deterministic counterparts. In interactive methods, the intermediate search results are presented to DMs to help them further understand the problem. Then, DMs can provide more preference information for guiding the search, and a desired solution which complies with DMs' preferences can be obtained. *A posteriori* methods imply that the optimization process is conducted without using DMs' preferences on uncertainties. A set of optimal solutions corresponding to different preferences are presented to DMs to select the most preferred solution based on their preferences. Generally, existing techniques on dealing with UOPs are either based on the knowledge of DMs' preferences on uncertainties or only suitable for limited types of DMs' preferences. These methods can deal with only one type of preference at one time and belong to *a priori* methods for UOPs. Actually, it is difficult for a DM to provide his/her preferences accurately, since the DM usually knows little about uncertainties before making decisions. Therefore, it is advisable to handle UOPs in an *a posteriori* manner.

In this chapter, an *a posteriori* decision-making framework of UOPs that covers several common uncertain models is first proposed. This *a posteriori* framework explicitly decomposes a UOP into a series of subproblems that stand for different preferences on handling uncertainties by associating each subproblem with a weight vector. Then, we synthesize the merits of the decomposition strategy and evolutionary optimization, and propose a subproblems co-solving evolutionary algorithm for UOPs, i.e., S-CoEA. It decomposes a UOP into a series of correlated deterministic subproblems based on different aggregation forms of the sorted sequence of sampled function values. Each subproblem is solved by using information from its neighboring subproblems. Besides, since the sampling of uncertain parameters is a very important factor in handling uncertainties, a sample-updating strategy based on historical information is proposed and used periodically for UOPs. In order to tackle the issue of mismatch induced by the sample-updating strategy, a solution-to-subproblem matching procedure is designed to place the nearest solution to each subproblem and is utilized after the sample-updating strategy. Finally, a subproblem-to-solution matching procedure is proposed to find a subproblem with the minimum constraint violation value for a newly generated solution.

4.1   Reformulation of UOPs under a Decomposition Framework

When optimizing UOPs, uncertain functions are unmeasurable. Therefore, in practice, an uncertain fitness function is often approximated by an aggregated value of random samples, such as the mean value, the worst-case value, and so on. Take the uncertain objective function $f(\mathbf{x}, \xi)$ as an example. If DMs want to make a neutral decision, the mean value $E\{f(\mathbf{x}, \xi)\}$ will be regarded as the objective function, and it can be approximated by an averaged sum of series of samples $\hat{f}(\mathbf{x}) = \sum_{i=1}^{n} \frac{1}{n} \cdot f(\mathbf{x}, \xi_i)$ [Liefooghe *et al.* (2007)], where $\xi_1, \xi_2, \ldots, \xi_n$ are n random samples of the uncertain parameter vector ξ. If DMs prefer to make a conservative decision, the worst-case value $\max\{f(\mathbf{x}, \xi)\}$ will be regarded as the objective function to be optimized. It is usually approximated by the worst sampled value over a number of samples $\hat{f}(\mathbf{x}) = \max_{i=1,\ldots,n} f(\mathbf{x}, \xi_i)$ [Li *et al.* (2016a); Xiong *et al.* (2017)]. When DMs' preferences on uncertainties are unavailable, different aggregation forms of sampled function values $f(\mathbf{x}, \xi_1), \ldots, f(\mathbf{x}, \xi_n)$ stand for different preferences on handling uncertainties and result in different uncertain models to be handled. Sort these sampled function values in an ascending order, and we can obtain $f(\mathbf{x}, \xi_{k_1}) \leq \cdots \leq f(\mathbf{x}, \xi_{k_n})$. Similarly, $g_j(\mathbf{x}, \xi_1), \ldots, g_j(\mathbf{x}, \xi_n)$ are samples of the jth uncertain constraint function g_j and $g_j(\mathbf{x}, \xi_{l_1}) \leq \cdots \leq g_j(\mathbf{x}, \xi_{l_n})$ is the sorted sequence for each constraint function $g_j (j = 1, \ldots, p)$. $f(\mathbf{x}, \xi_{k_1}), \ldots, f(\mathbf{x}, \xi_{k_n})$ and $g_j(\mathbf{x}, \xi_{l_1}), \ldots, g_j(\mathbf{x}, \xi_{l_n})$ are order statistics,[1] which are the most fundamental tools in statistics. In statistics, the order statistic is sufficient,[2] which means it contains all the information needed to compute any estimate of a parameter [Schervish (2012)].

Based on the above, an *a posteriori* decision-making framework that covers multiple uncertain models corresponding to the UOP *P0* is formulated as follows:

$$P1 : \text{minimize} \quad \sum_{i=1}^{n} w_{f,i} \cdot f(\mathbf{x}, \xi_{k_i})$$

$$\text{subject to} \quad \sum_{i=1}^{n} w_{g_j,i} \cdot g_j(\mathbf{x}, \xi_{l_i}) \leq 0, \quad j = 1, \ldots, p, \tag{4.1}$$

where $\mathbf{w}_f = [w_{f,1}, \ldots, w_{f,n}](\sum_{i=1}^{n} w_{f,i} = 1)$ and $\mathbf{w}_{g_j} = [w_{g_j,1}, \ldots, w_{g_j,n}]$ $(\sum_{i=1}^{n} w_{g_j,i} = 1, j = 1, \ldots, p)$ are weight vectors of the objective function f and the jth constraint function $g_j (j = 1, \ldots, p)$, respectively. According to these notations, different weight vectors $\mathbf{w}_f$ and $\mathbf{w}_{g_j}$ correspond to different uncertain models.[3] An illustrative example of the *a posteriori* framework for uncertain function $f(\mathbf{x}, \xi)$ is shown in Fig. 4.1 where four common uncertain models with different weight vectors are illustrated.

Given the above description, the model *P1* can cover DMs' common preferences on uncertainties. Hence, a UOP can be decomposed into a series of subproblems by assigning a weight vector to each subproblem. The neighborhood relations among subproblems are defined according to the distances between their weight vectors. Usually, it can be expected that optimal solutions of two neighboring subproblems are very similar. Each subproblem will be optimized by using information from its neighboring subproblems, and all subproblems can be solved simultaneously in a single run. Thus, a UOP can be optimized via co-solving a series of correlated subproblems by using the neighbor information.

---

[1] Given any random variables $X_1, X_2, \ldots, X_n$, the order statistics $X_{(1)}, X_{(2)}, \ldots, X_{(n)}$ are random variables that are defined by sorting the values (realizations) of $X_1, X_2, \ldots, X_n$ in increasing order [Schervish (2012)].

[2] A statistic is sufficient with respect to a statistical model and its associated unknown parameter if no other statistic that can be calculated from the same sample provides any additional information as to the value of the parameter.

[3] The mean-case value, best-case value, worst-case value and median-case value of the uncertain function $f(\mathbf{x}, \xi)$ can be approximated by $\sum_{i=1}^{n} w_{f,i} \cdot f(\mathbf{x}, \xi_i)$ when $w_{f,i} = 1/n (i = 1, \ldots, n)$; $w_{f,1} = 1, w_{f,i} = 0 (i = 2, \ldots, n)$; $w_{f,n} = 1, w_{f,i} = 0 (i = 1, \ldots, n-1)$; and $w_{f,k} = 1, w_{f,i} = 0 \left(i = 1, \ldots, n, i \neq k, k = \begin{cases} n/2, & n \text{ is even} \\ (n+1)/2, & n \text{ is odd} \end{cases} \right)$, respectively.

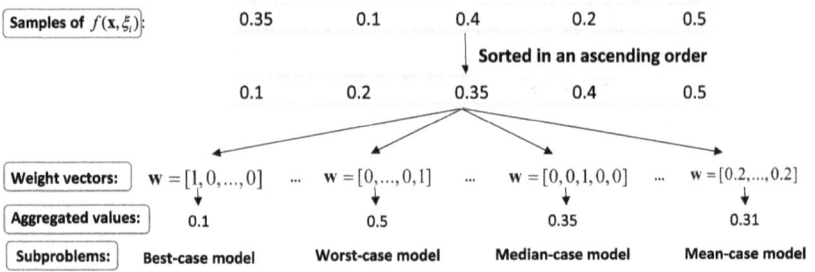

Fig. 4.1   Illustration of the *a posteriori* framework for uncertain function $f(\mathbf{x}, \xi)$.

4.2   Framework of S-CoEA

4.2.1   *Algorithmic Framework*

S-CoEA converts a UOP into N subproblems and optimizes them synergistically in a single run. Let $\{\mathbf{w}^1, \ldots, \mathbf{w}^N\}$ be a set of evenly spread weight vectors, and the neighborhood of weight vector $\mathbf{w}^i$ is defined as the set of its several closest weight vectors in $\{\mathbf{w}^1, \ldots, \mathbf{w}^N\}$. The neighborhood of the ith subproblem consists of all subproblems with the weight vectors from the neighborhood of $\mathbf{w}^i$ and is denoted as $B(i)$. Each weight vector $\mathbf{w}^i = [\mathbf{w}_f^i, \mathbf{w}_{g_1}^i, \ldots, \mathbf{w}_{g_p}^i]$ consists of an objective's weight vector $\mathbf{w}_f^i$ and p constraint objectives' weight vectors $[\mathbf{w}_{g_1}^i, \ldots, \mathbf{w}_{g_p}^i]$. When optimizing each subproblem, a standard EA with the feasibility rule [Deb (2000)] is adopted.

During the search procedure, S-CoEA requires a set of parameters as input, including:

- N: The number of weight vectors, which is the same as the population size;
- n: Sample size for each solution;
- T: Neighborhood size;
- δ: Probability of selecting mate solutions from its neighborhood;
- n_r: Maximum number of replacement when updating neighborhood subproblems;
- I: Iteration interval of updating samples of uncertain parameters; and
- NFE: Maximum number of function evaluations.

The algorithmic description of S-CoEA is presented in **Algorithm 9**, in which *rand* means a random number in $[0, 1]$.

Algorithm 9 Framework of S-CoEA

Require: A UOP, related parameters.

Ensure: An evolving population P.

1: Initialize N evenly spread weight vectors.

2: **for** $i = 1$ to N **do**

3:　　Set the neighborhood of the ith subproblem $B(i)$.

4: **end for**

5: Initialize samples of uncertain parameters randomly.

6: Randomly initialize the evolving population $P = \{\mathbf{x}^1, \ldots, \mathbf{x}^N\}$; calculate sampled values of the objective function $\mathbf{f}^i = [f_1^i, \ldots, f_n^i]$ and constraint functions $\mathbf{g}_j^i = [g_{j,1}^i, \ldots, g_{j,n}^i](j = 1, \ldots, p)$; set $gen = 0, count = n \cdot N$.

7: Use the solution-to-subproblem matching procedure (**Algorithm 11**) to match solutions with subproblems.

8: **while** $count \leq NFE$ **do**

9:　　**if** gen is a multiple of I **then**

10:　　　　Update the samples of uncertain parameters by applying the sample-updating scheme (**Algorithm 10**).

11:　　　　$count = count + n \cdot N$.

12:　　　　Use the solution-to-subproblem matching procedure (**Algorithm 11**) to match solutions with subproblems.

13:　　**end if**

14:　　**for** $i = 1$ to N **do**

15:　　　　$P = \begin{cases} B(i), & \text{if } rand < \delta \\ \{1, 2, \ldots, N\}, & otherwise \end{cases}$

16:　　　　Select parent individuals from P randomly and apply a certain reproduction operator to generate a new solution $\mathbf{y}$. If $\mathbf{y}$ is infeasible, repair it.

17:　　　　$count = count + n$.

18:　　　　Use the subproblem-to-solution matching procedure (**Algorithm 12**) to find a subproblem k for $\mathbf{y}$.

19:　　　　Compare $\mathbf{y}$ with neighboring solutions of the subproblem k and update these neighboring solutions by comparing objective values or using the feasibility rule if the subproblem is a constrained one.

20:　　**end for**

21:　　$gen = gen + 1$.

22: **end while**

4.2.2   Generation of Weight Vectors $\{\mathbf{w}^1, \ldots, \mathbf{w}^N\}$

A structured set of weight vectors $\mathbf{w}^k = [w_1^k, \ldots, w_n^k](k = 1, \ldots, N)$ are generated by taking each individual weight coefficient as a value from $\{\frac{0}{H}, \frac{1}{H}, \ldots, \frac{H}{H}\}$ and satisfying the normality constraint $\sum_{i=1}^{n} w_i^k = 1(k = 1, \ldots, N)$, where H is a controllable parameter. For UOPs with uncertainties in both objective and constraint functions, each set of weight vectors should be generated according to the above-mentioned procedure for each uncertain function, and the combinations of these sets of weight vectors are regarded as final weight vectors. Take a UOP with an uncertain objective function f and one uncertain constraint function g as an example. Suppose $\{\mathbf{p}^1, \ldots, \mathbf{p}^{N_1}\}$ and $\{\mathbf{q}^1, \ldots, \mathbf{q}^{N_2}\}$ are two sets of weight vectors generated for the objective function and the uncertain constraint function, respectively. Then, a set of final weight vectors is $\{\mathbf{w}^1, \ldots, \mathbf{w}^k, \ldots, \mathbf{w}^{N_1 \times N_2}\}$, where $\mathbf{w}^k = [\mathbf{p}^i, \mathbf{q}^j](i = 1, \ldots, N_1; \ j = 1, \ldots, N_2; k = 1, \ldots, N_1 \times N_2)$. The process of generating weight vectors is given in Fig. 4.2.

4.2.3   Sample-updating Scheme

The sampling strategy is used to alleviate the detrimental effects of uncertainties, and fixed samples are not favorable for gathering landscape information of uncertain functions and usually mislead the evolutionary process. Therefore, a sample-updating procedure based on historical knowledge is proposed and used periodically for UOPs.

Firstly, n samples of the uncertain parameter ξ are newly generated and the sampled values of the objective function $\mathbf{f}^i$ and constraint functions $\mathbf{g}_j^i (j = 1, \ldots, p)$ for each solution $\mathbf{x}^i (i = 1, \ldots, N)$ are calculated as described in line 1 of **Algorithm 10**. The union of the original sample set S_1 and the newly generated sample set S_2 is denoted as $S = S_1 \cup S_2$ whose size is $2n$. Then, we collect information of the objective function and p constraint functions across N solutions for each sample as described in lines 2–5 of **Algorithm 10**. Next, n representative samples are selected

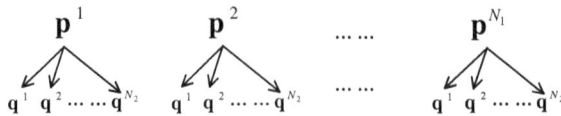

Fig. 4.2   Illustration of generating weight vectors for UOPs with one uncertain constraint function ($\{\mathbf{p}^1, \ldots, \mathbf{p}^{N_1}\}$ and $\{\mathbf{q}^1, \ldots, \mathbf{q}^{N_2}\}$ are two sets of weight vectors generated for the objective function and the uncertain constraint function, respectively).

Algorithm 10 Sample-updating Scheme

Require: N solutions $(\mathbf{x}^1, \mathbf{f}^1, \mathbf{g}_1^1, \ldots, \mathbf{g}_p^1), \ldots, (\mathbf{x}^N, \mathbf{f}^N, \mathbf{g}_1^N, \ldots, \mathbf{g}_p^N)$, where $\mathbf{f}^i = [f_1^i, \ldots, f_n^i]$ and $\mathbf{g}_j^i = [g_{j,1}^i, \ldots, g_{j,n}^i](j = 1, \ldots, p)$. $f_1^i, \ldots, f_n^i$ and $g_{j,1}^i, \ldots, g_{j,n}^i, (j = 1, \ldots, p)$ are samples of the uncertain objective function and the jth uncertain constraint function, respectively. Sample size n and the set of original samples $S_1 = \{\xi_1, \ldots, \xi_n\}$.

Ensure: Updated sample set S' and updated N solutions.

1: Randomly sample the uncertain parameter ξ for n times and get the newly generated sample set $S_2 = \{\xi_{n+1}, \ldots, \xi_{2n}\}$; calculate sampled values of the objective function $\mathbf{f}^i = [f_{n+1}^i, \ldots, f_{2n}^i]$ and constraint functions $\mathbf{g}_j^i = [g_{j,(n+1)}^i, \ldots, g_{j,2n}^i](j = 1, \ldots, p)$ for each solution $\mathbf{x}^i(i = 1, \ldots, N)$; set $S = S_1 \cup S_2$.

2: **for** $k = 1$ to $2n$ **do**

3:   $obj_k = \sum_{i=1}^{N} f_k^i$; $cv_{j,k} = \sum_{i=1}^{N} g_{j,k}^i (j = 1, \ldots, p)$.

4:   $p_k = (obj_k, cv_{1,k}, \ldots, cv_{p,k})$. //Accumulate information of the objective function and p constraint functions across all N solutions for each sample.

5: **end for**

6: Use the farthest-candidate approach [Chen *et al.* (2015)] to select a set of n representative samples, denoted as S', from S.

7: Update objective function values and p constraint function values of N solutions.

---

from the set of all samples S by using the farthest-candidate approach [Chen *et al.* (2015)]. Suppose that we are going to select n best points from $2n$ points. Whenever a new point is to be selected, the candidate point in the unselected points that are farthest from the selected points will be selected.

4.2.4 Solution-to-Subproblem Matching Procedure

After the sample-updating procedure is employed, a solution which is good for the current subproblem may no longer perform well since the fitness values of the uncertain objective function and uncertain constraint functions of this subproblem have been changed. Thus, a solution-to-subproblem matching procedure is proposed. In this matching procedure, the solution that has the minimum distance value to a certain subproblem among N current solutions is matched with this subproblem. The distance value from the solution

---

Algorithm 11 Solution-to-Subproblem Matching

Require: N solutions $(\mathbf{x}^1, \mathbf{f}^1, \mathbf{g}_1^1, \ldots, \mathbf{g}_p^1), \ldots, (\mathbf{x}^N, \mathbf{f}^N, \mathbf{g}_1^N, \ldots, \mathbf{g}_p^N)$ and N
    subproblems with weight vectors $\mathbf{w}^1, \ldots, \mathbf{w}^N$.
Ensure: Matched pairs $(\mathbf{x}^k, \mathbf{f}^k, \mathbf{g}_1^k, \ldots, \mathbf{g}_p^k) \sim \mathbf{w}^l (k, l = 1, \ldots, N)$.
  1: Initialize $\Omega = \{1, 2, \ldots, N\}$.
  2: **while** Ω is nonempty **do**
  3:    Randomly select a weight vector $\mathbf{w}^l, l \in \Omega$.
  4:    **for** $i = 1$ to N **do**
  5:      Calculate the distance d_i^l according to Eq. (4.2).
  6:    **end for**
  7:    $k = \mathrm{argmin}_{i \in \{1, \ldots, N\}} d_i^l$.
  8:    $\mathbf{f}^k = \inf, \mathbf{g}_1^k = \inf, \ldots, \mathbf{g}_p^k = \inf; \Omega = \Omega \backslash \{l\}$.
  9: **end while**

$(\mathbf{x}^i, \mathbf{f}^i, \mathbf{g}_1^i, \ldots, \mathbf{g}_p^i)$ to the subproblem with $\mathbf{w}^l$ is defined as:

$$
d_i^l = \begin{cases} \displaystyle\sum_{j=1}^n w_j^l \cdot f_j^i, & \text{UOPs without uncertain constraints,} \\[2em] \displaystyle\sum_{k=1}^p \sum_{j=1}^n w_{g_k,j}^l \cdot |g_{k,j}^i|, & \text{UOPs with uncertain constraints.} \end{cases} \tag{4.2}
$$

    Specifically, for UOPs without uncertainties in constraints, the solution that has the minimum objective value to a certain subproblem among all N current solutions is matched with this subproblem. For UOPs with uncertain constraints, the solution-to-subproblem matching procedure involves finding the solution that has the minimum distance to the subproblem among N current solutions for a subproblem. This matching procedure will be used after the sample-updating scheme. It can place the nearest solution to each subproblem at a large degree, which is beneficial to convergence.

4.2.5 *Subproblem-to-Solution Matching Procedure*

When a new solution is generated, it may perform badly for the current subproblem but perform well for another subproblem. In order to avoid wasting potentially useful solutions and make the best use of them, the subproblem-to-solution matching procedure is proposed to find a subproblem with the minimum constraint violation value for the new solution. The constraint violation value of the new solution $\mathbf{y}$ regarding the subproblem

---

Algorithm 12 Subproblem-to-Solution Matching

Require: New generated solution $(\mathbf{y}, \mathbf{f}^y, \mathbf{g}_1^y, \ldots, \mathbf{g}_p^y)$ and N subproblems with weight vectors $\mathbf{w}^1, \ldots, \mathbf{w}^N$.
Ensure: The index of the selected subproblem k.
1: **for** $l = 1$ to N **do**
2:    Calculate the constraint violation c^l according to Eq. (4.3).
3: **end for**
4: $k = \mathrm{argmin}_{i \in \{1, \ldots, N\}} c^i$.

---

with $\mathbf{w}^l$ is defined as:

$$
c^l = \begin{cases}
\displaystyle\sum_{j=1}^{n} w_j^l \cdot f_j^y, & \text{UOPs without uncertain constraints,} \\[2ex]
\displaystyle\sum_{k=1}^{p}\sum_{j=1}^{n} w_{g_k,j}^l \cdot \max(g_{k,j}^y, 0), & \text{UOPs with uncertain constraints.}
\end{cases}
$$

$$(4.3)$$

For UOPs without uncertainties in constraints, the subproblem with the minimum objective value of the newly generated solution among all N subproblems is matched with the newly generated solution. As to UOPs with uncertain constraints, the subproblem that has the minimum constraint violation value is selected for a newly generated solution. This matching procedure will be used after the generation of a new solution $\mathbf{y}$, which is beneficial to convergence.

4.2.6 *Discussions*

4.2.6.1 *Main differences between S-CoEA and existing approaches for UOPs*

The proposed S-CoEA and existing uncertain algorithms are designed for UOPs, but they handle UOPs in totally different ways. Specifically, existing algorithms either transform a UOP into its deterministic counterpart by using DMs' preferences on uncertainties or propose special mechanisms to handle uncertainties. They usually can only deal with one type of DMs' preferences on uncertainties, i.e., one uncertain model. However, S-CoEA can handle a series of uncertain models collaboratively without using DMs' preferences. A set of optimal solutions obtained from S-CoEA will be presented to a DM, and the DM can choose one according to his/her preferences.

Table 4.1   Time Complexity Analysis of S-CoEA.

Procedure	Worst-case Time Complexity
Initialization	$O((n + n \cdot p) \cdot N)$
Sample-updating	$O((n + n \cdot p) \cdot (4n + N))$
Solution-to-subproblem matching	$O((n + n \cdot p) \cdot N)$
Generate a new solution	$O(n + n \cdot p)$
Subproblem-to-solution matching	$O(n + n \cdot p)$
Update neighborhood solutions	$O((n + n \cdot p) \cdot T)$

4.2.6.2   *Main differences between S-CoEA and decomposition-based MOEAs*

Both S-CoEA and decomposition-based multi-objective evolutionary algorithms (MOEAs) introduce the concept of decomposition and decompose the original problem into a collection of subproblems. The neighborhood relations among these subproblems are defined according to the Euclidean distance between their weight vectors. N subproblems are co-solved by using the neighboring information in parallel. However, S-CoEA aims for the UOPs and decomposition-based MOEAs are designed for MOPs. Besides, optimal solutions obtained via S-CoEA are irrelevant to the concept of Pareto dominance, and they are just different optima for different uncertain models.

4.2.6.3   *Computational complexity analysis*

The time complexity analysis of S-CoEA is presented in Table 4.1. Since $T \leq N$, the worst-case time complexity of S-CoEA is $O((n+n \cdot p) \cdot (4n+N))$. It should be noted that with the same amount of computational resources, the optimal solutions of different uncertain models are obtained by using S-CoEA. In contrast, standard EAs can only get the optimal solution of a single uncertain model.

4.3   Experimental Design

4.3.1   *Uncertain Continuous Test Instances*

Up to now, there are no widely accepted test problems for UOPs. In noisy UOPs, noise with different strength levels is implemented as an additive perturbation on well-known deterministic benchmark problems. The majority of studies on robust UOPs concentrate on specific optimization problems, such as job shop scheduling and design optimization [Jin and Branke (2005)].

A set of benchmarks of UOPs usually includes different types of uncertainties (e.g., normal distributions, uniform distributions, and so on), the uncertain objective or uncertain constraint functions, and the different forms that uncertainties are introduced into uncertain functions (e.g., decision variables and environmental variables) [Beyer and Sendhoff (2007)]. A set of 7 robust UOPs and a noisy one with various characteristics and different levels of difficulties are collected from the literature. The formulations and characteristics of these functions are summarized in Table 4.2.

The quadratic N-dimensional sphere $f(x) = \sum_{i=1}^{N} x_i^2$ is a simple scalable test function. It was firstly considered in the context of perturbation induced in design variables in [Beyer *et al.* (2003)]. Thus, the quadratic sphere model is taken as an instance of UOPs and named as *Test1*. *Test1*, with two types of uncertainties, i.e., uniform and normal distributions, is considered and denoted as *Test1-U* and *Test1-N*, respectively. The sphere model with environmental uncertainties, named *Test2*, also serves as a test instance of UOPs. According to [Beyer and Sendhoff (2006)], *Test2* instances with different levels of uncertainties have different optimal solutions.[4] Therefore, three uniform uncertainties with different strength levels and a normal one are considered. *Test2-U1*, *Test2-U2*, *Test2-U3* and *Test2-N* represent *Test2* with uniform uncertainties of three different strength levels and normal uncertainties. The test problem *EVM* was used in [Liu (2009)] and modeled as an expected value model. *EVM* is characterized by different types of uncertainties.

A new class of test functions that were motivated from the design optimization of gas-turbine blades has been proposed in [Sendhoff *et al.* (2002)]. Unlike these functions considered so far, the newly proposed function class exhibits a change of the mean-case landscape depending on the uncertain parameter ε. These instances are termed as functions with noise-induced multimodality (*FNIMs*), and 4 variants of *FNIMs* are considered.

To be specific, *FNIM_f0* is characterized by additive noise and is actually not an *FNIM*. However, the evolutionary strategy exhibits a similar behavior on *FNIM_f0*. Here it is selected as an example of noisy UOPs. In *FNIM_f2*, uncertainties enter the system via decision variables. The uniform and normal uncertainties with two different levels of uncertainties

---

[4]For $|\xi| \leq \varepsilon$, when $0 \leq \varepsilon < 1$, the optimal solution is $x = 0$; when $\varepsilon > 1$ all x which fulfill $\|x\| = b^{1/\beta}$ are optimal; and when $\varepsilon = 1$, all x which fulfill $\|x\| < b^{1/\beta}$ are optimal. When assuming $\xi \sim N(0, \varepsilon^2)$, the standard deviation ε has no influence on the location of the optimum when optimizing the mean value of *Test2* [Beyer and Sendhoff (2006)].

Table 4.2   List of Continuous Test Instances of UOPs.

Name	Problem Formulation	Uncertainties	Characteristics
Test1 [Beyer et al. (2003)]	$f(x,\xi) = \sum_{i=1}^{10}(x_i+\xi_i)^2, \ -1 \leq x_i \leq 1$	U: $\xi_i \sim U(-0.1, 0.1)$; N: $\xi_i \sim N(0, 0.1)$.	Induced by decision variables.
Test2 [Beyer and Sendhoff (2006)]	$f(x,\xi) = a + (\xi+1)\cdot \|x\|^\beta - b\xi,$ $a=-5, \beta=1, b=-1, -1 \leq x_i \leq 1, i=1,\dots,10$	U1: $\xi \sim U(-0.5, 0.5)$; U2: $\xi \sim U(-1, 1)$; U3: $\xi \sim U(-1.5, 1.5)$; N: $\xi \sim N(0, 0.1)$.	Induced by environmental variables.
EVM [Liu (2009)]	$f(x,\xi) = \sqrt{\sum_{i=1}^{3}(x_i-\xi_i)^2}$ $s.t. \ \sum_{i=1}^{3} x_i^2 \leq 10$	$\xi_1 \sim U(1,2), \xi_2 \sim N(3,1),$ $\xi_3 \sim Exp(4)$.	Induced by environmental variables and different types of uncertainties.
FNIM-f0 [Sendhoff et al. (2002)]	$f(x,\xi) = -5 + \frac{\sum_{i=1}^{N-1} x_i^2}{x_N^2+1} + x_N^2 + \xi, N=40$	$\xi \sim N(0, 0.5)$.	Additive noise.
FNIM-f2 [Sendhoff et al. (2002)]	$f(x,\xi) = -5 + \frac{(x_{N-1}+\xi)^2+\sum_{i=1}^{N-2} x_i^2}{x_N^2+b} + x_N^2$ $b=1, N=40$	U1: $\xi \sim U(-0.5, 0.5)$; U2: $\xi \sim U(-1.5, 1.5)$; N1: $\xi \sim N(0, 0.5)$; N2: $\xi \sim N(0, 1.5)$.	Functions with noise-induced multimodality.
FNIM-f3 [Sendhoff et al. (2002)]	$f(x,\xi) = -5 + \frac{\sum_{i=1}^{N_1-1} x_i^2 + \sum_{i=N_1}^{N_2-1}(x_i+\xi_i)^2}{\sum_{i=N_2}^{N} x_i^2+b} + x_N^2$ $b=1, N=40, N_1=23, N_2=39$	U1: $\xi \sim U(-0.25, 0.25)$; U2: $\xi \sim U(-1, 1)$; N1: $\xi \sim N(0, 0.25)$; N2: $\xi \sim N(0, 0.1)$.	Functions with noise-induced multimodality.

Table 4.2 *(Continued)*

Name	Problem Formulation	Uncertainties	Characteristics
FNIM-f4 [Beyer and Sendhoff (2006)]	$f(x,\xi) = -5 + \dfrac{\sum_{i=1}^{N-1}(x_i+\xi_i)^2}{b+x_N^2} + x_N^2$ $b = 1, N = 40$	U1: $\xi \sim U(-0.15, 0.15)$; U2: $\xi \sim U(-1, 1)$; N1: $\xi \sim N(0, 0.15)$; N2: $\xi \sim N(0, 1)$.	Functions with noise-induced multimodality.
Interval1 [Jiang et al. (2008)]	$f(x,\xi) = \xi_1(x_1-2)^2 + \xi_2(x_2-1)^2 + \xi_3 x_3$ $5 \le \xi_1 x_1^2 - \xi_2^2 x_2 + \xi_3 x_3 \le 8$ $s.t.\ 1.2 \cdot x_1 + x_2 + 1.3^2 \cdot x_3^2 + 1 \ge 15$ $-1 \le x_1 \le 5,\ -3 \le x_2 \le 6,\ -1 \le x_3 \le 7$	$\xi_1 \sim U(1, 1.3)$, $\xi_2 \sim U(0.9, 1.1)$, $\xi_3 \sim U(1.2, 1.4)$.	Uncertainties exist in the first constraint function.
Interval2 [Jiang et al. (2008)]	$f(x,\xi) = \xi_1(x_1-2)^2 + \xi_2(x_2-1)^2 + \xi_3 x_3$ $5 \le 1.2 \cdot x_1^2 - x_2 + 1.3 \cdot x_3 \le 8$ $s.t.\ \xi_1 x_1 + \xi_2 x_2 + \xi_3^2 x_3^2 + 1 \ge 15$ $-1 \le x_1 \le 5,\ -3 \le x_2 \le 6,\ -1 \le x_3 \le 7$	$\xi_1 \sim U(1, 1.3)$, $\xi_2 \sim U(0.9, 1.1)$, $\xi_3 \sim U(1.2, 1.4)$.	Uncertainties exist in the second constraint function.
Interval3 [Jiang et al. (2008)]	$f(x,\xi) = \xi_1(x_1-2)^2 + \xi_2(x_2-1)^2 + \xi_3 x_3$ $5 \le \xi_1 x_1^2 - \xi_2^2 x_2 + \xi_3 x_3 \le 8$ $s.t.\ \xi_1 x_1 + \xi_2 x_2 + \xi_3^2 x_3^2 + 1 \ge 15$ $-1 \le x_1 \le 5,\ -3 \le x_2 \le 6,\ -1 \le x_3 \le 7$	$\xi_1 \sim U(1, 1.3)$, $\xi_2 \sim U(0.9, 1.1)$, $\xi_3 \sim U(1.2, 1.4)$.	Uncertainties exist in two constraint functions.

are taken into consideration.[5] *FNIM_f2-U1*, *FNIM_f2-U2*, *FNIM_f2-N1* and *FNIM_f2-N2* represent *FNIM_f2* with uniform and normal uncertainties of two different levels. A generalization of *FNIM_f2*, named *FNIM_f3*, has been presented in [Sendhoff *et al.* (2002)] with the consideration of providing deeper insight in certain aspects of robust optimization. If the uncertainty strength ε exceeds a threshold, the robust counterpart of *FNIM_f3* changes from a unimodal function to a multi-modal one. *FNIM_f3* with uniform and normal uncertainties of two different levels are taken into consideration and denoted as *FNIM_f3-U1*, *FNIM_f3-U2*, *FNIM_f3-N1* and *FNIM_f3-N2*, respectively. *FNIM_f4* was proposed in [Beyer and Sendhoff (2006)] to predict the behavior of (μ, λ)-ES. *FNIM_f4-U1*, *FNIM_f4-U2*, *FNIM_f4-N1* and *FNIM_f4-N2* represent *FNIM_f4* test instances with uniform and normal uncertainties of two different levels.[6]

The set of test instance *Interval* comes from [Jiang *et al.* (2008)]. Different from the above test instances, it is featured by uncertain constraint functions. Variants of *Interval*, namely *Interval1*, *Interval2* and *Interval3*, are designed in this chapter.

4.3.2   *Uncertain Discrete Test Instances*

The unrepairable series-parallel standby redundancy system [Taboada *et al.* (2007); Ji *et al.* (2019)] is taken as an example in this paper, and the following assumptions are used:

- Each component has only two possible states: functioning or failed;
- The states of all components are statistically independent;
- All the components in a subsystem are identical.

Under the above-mentioned assumptions, the formulation of RAPs for a series-parallel system is defined as:

$$\text{maximize}\quad f(x, \xi) = \prod_{i=1}^{m}\left(1 - \prod_{j=1}^{n_i}(1 - r_{ij}(\xi_{ij}))^{x_{ij}}\right)$$

$$\text{subject to}\quad \sum_{i=1}^{m}\sum_{j=1}^{n_i}c_{ij}x_{ij} \leq c \quad 1 \leq \sum_{j=1}^{n_i}x_{ij} \leq n_{\max,i}, \quad i = 1,\ldots,m \tag{4.4}$$

---

[5]Assuming $\xi \sim U(-\varepsilon, \varepsilon)$ or $\xi \sim N(0, \varepsilon^2)$, different optimal solutions will be obtained for *FNIM_f2* with $\varepsilon > b$ and $\varepsilon \leq b$ [Sendhoff *et al.* (2002)].
[6]Different optimal solutions will be obtained for *FNIM_f4-U1* with $\varepsilon \leq b/\sqrt{N-1}$ and $\varepsilon > b/\sqrt{N-1}$ [Beyer and Sendhoff (2006)].

where x_{ij} is the decision variable which represents the quantity of component j used in sub-system i. m is the number of sub-systems, and n_i is the number of available component choices for sub-system i. c_{ij} is the cost of component j in sub-system i. $r_{ij}(\xi_{ij})$ represents the uncertain reliability of component j in sub-system i and depends on a random parameter $\xi_{ij}(i = 1, \ldots, m, j = 1, \ldots, n_i)$. Different distributions of the reliability value $r_{ij}(\xi_{ij})$ can be available through expert knowledge before the system design. Besides, there is a constraint on the maximum number of components for each sub-system. That is, for each sub-system i, the maximum number of components which can be parallelized is $n_{max,i}$.

For RAPs, two well-known benchmarks which consist of 3 sub-systems with 5 components and 14 subsystems with 4 components [Khalili-Damghani and Amiri (2012); Zhang *et al.* (2017)] are considered. The two well-known benchmark cases, denoted as RAP1 and RAP2, will be regarded as bases for generating uncertain test instances.

As to uncertain RAPs, the probability distribution that the uncertain reliability may follow can be applied over the deterministic instances. Specifically, uncertain parameters ξ are sampled from the following uniform probability distribution:

- Uniform distribution: $r_{ij}(\xi) \sim U((1 - \alpha) \cdot r_{ij}, (1 + \alpha) \cdot r_{ij})$

The central tendency of the distribution always corresponds to the deterministic reliability value r_{ij} of the instance under consideration. The parameter α is used to tune the degree of the deviation of uncertainties. In the following, two α-values, namely $\alpha = 0.1, 0.2$, will be considered. For convenience of results display, we denote the uncertain scenarios with uncertain parameters following a uniform distribution with $\alpha = 0.1$ and $\alpha = 0.2$ by *U1* and *U2*, respectively.

4.3.3   *Performance Measures*

Evaluation of optimal solutions of deterministic problems is straightforward by utilizing their objective function values. However, the objective function of a UOP is stochastic and unmeasurable. Multiple repeated evaluations of an optimum of a UOP are a sequence of stochastic values, thus it is reasonable to make use of statistical hypothesis tests when comparing different optima obtained via different approaches for a UOP. Two commonly used statistical hypothesis tests, i.e., Mann–Whitney U test and Kolmogorov–Smirnov test, are employed to evaluate the performance of all compared algorithms.

Both the Mann–Whitney U test and Kolmogorov–Smirnov test are non-parametric tests that are employed with respect to ordinal data. The Mann–Whitney U test is used to test whether two independent samples represent two populations with different median values. One reason for employing the Mann–Whitney U test is its ability of reducing or eliminating impacts of outliers. If the result of the Mann–Whitney U test is significant, it indicates that there is a significant difference between the medians of two samples. The Kolmogorov–Smirnov test is used for determining whether two independent samples are drawn from the same distribution. It is sensitive to differences in both location and shape of the empirical cumulative distribution functions of the two samples. If there is a significant difference at any point along the two cumulative frequency distributions, we can conclude that there is a high likelihood the samples are derived from different populations. The two statistical hypothesis tests are conducted at a 5% significance level to test the significance of differences between the optimal solutions yielded by S-CoEA and its competitors.

4.3.4   *Parameter Settings*

For a fair comparison, the choice of parameters that S-CoEA and comparison algorithms share keeps the same. Specifically, the number of samples is set to $n = 5$, and the population size is set as $N = 86$[7] for all test instances except for three *Interval* test instances. As to *Interval1* and *Interval2*, the population size is $N = 121$.[8] For *Interval3*, the population size is $N = 216$.[9] The number of function evaluations is set as $NFE = 50,000$ for all test

---

[7]For UOPs without uncertain constraints, the number of weight vectors is determined by $N = C_{H+n-1}^{n}$ (n is the number of samples and H is a controlled parameter). In order to set a reasonable population size, the 2-layer weight vector generation method proposed in [Li *et al.* (2015b)] is employed. Here we set 2 H values, i.e., $H_1 = 4, H_2 = 2$. Besides, an extra weight vector whose each element equals, i.e., $w = [1/n, \ldots, 1/n]$, is added. Thus, the population size is $N = C_{4+5-1}^{5-1} + C_{2+5-1}^{5-1} + 1 = 70 + 15 + 1 = 86$.

[8]For UOPs with one uncertain constraint, in order to set a reasonable population size, 2 sets of 11 weight vectors are generated by using the two-layer generation method [Li *et al.* (2015b)] with 2 H values, i.e., $H_1 = 1, H_2 = 1$. Then, final weight vectors are the combination of the two sets of weight vectors, and the number of final weight vectors is $N = (C_{1+5-1}^{5-1} + C_{1+5-1}^{5-1} + 1) \cdot (C_{1+5-1}^{5-1} + C_{1+5-1}^{5-1} + 1) = 11^2 = 121$.

[9]For UOPs with 2 uncertain constraints, in order to set a reasonable population size, 3 sets of 11 weight vectors are generated by using the Das and Dennis's method [Das and Dennis (1998)] with $H = 1$. Thus, final weight vectors are the combination of the three sets of weight vectors, and the number of final weight vectors is $N = (C_{1+5-1}^{5-1} + 1) \cdot (C_{1+5-1}^{5-1} + 1) \cdot (C_{1+5-1}^{5-1} + 1) = 6^3 = 216$.

problems, and all compared algorithms stop when the number of function evaluations reaches the maximum number. Besides, for the continuous test insatces, the DE/rand/1/bin operator with $CR = 0.9, F = 0.5$ [Rakshit et al. (2014)] and the Gaussian mutation [Jiang et al. (2008)] are adopted. As to the discrete test instances, the two-point crossover, one-point mutation, and random repair mechanism are used for generating a feasible new solution. Finally, each algorithm is executed 30 times independently on each instance.

In addition to the above-mentioned common parameters, parameter settings in S-CoEA and its variants include: the neighborhood size is set as $T = \lfloor 0.9N \rfloor$; the probability of selecting mate solutions from the neighborhood is set as $\delta = 0.9$; the maximal number of replacement is set as $n_r = \lfloor 0.1N \rfloor$; and the iteration interval of utilizing the sample-updating strategy is set as $I = 10$.

4.4   Detailed Analysis on Behavior of S-CoEA

This part of the experiment is designed to study effects of the solution-to-subproblem matching procedure and the subproblem-to-solution matching procedure. When evaluating the performance of different algorithms, a set of 10,000 samples of uncertain parameters are randomly generated for each test instance. Then, the optimal solution obtained by a certain algorithm will be evaluated using the set of high density samples, and the sequence of stochastic objective values will be used to perform two statistical hypothesis tests.

As mentioned above, the solution-to-subproblem matching procedure and the subproblem-to-solution matching procedure are both beneficial for convergence. For UOPs, a solution that is not suitable for the current subproblem may be good for another subproblem. In order to make the best use of computational resources, the solution-to-subproblem matching strategy is needed for improving the algorithm performance. What's more, when a new solution is generated, the subproblem-to-solution matching procedure is proposed to further enhance the convergence of the algorithm.

Do the above two matching mechanisms indeed play an important role in S-CoEA? In order to answer this question, three S-CoEA variants, denoted as S-CoEA_No_No, S-CoEA_No_CV and S-CoEA_D_No, are developed for comparison with the original S-CoEA. Detailed descriptions of the three variants will be given in the following. Furthermore, for UOPs with uncertain constraints, why is the distance between a subproblem and a solution selected as the matching criterion for the solution-to-subproblem matching procedure? Why is the constraint violation of a solution regarding a subproblem adopted as the matching criterion for the subproblem-to-solution matching procedure? To illustrate the effectiveness of the two criteria, another three variants of S-CoEA, including S-CoEA_D_D, S-CoEA_CV_D and S-CoEA_CV_CV, are designed.

S-CoEA_No_No: Different from S-CoEA, the two matching procedures are both removed.

S-CoEA_No_CV: Different from S-CoEA, the solution-to-subproblem matching procedure is removed. The constraint violation value is still adopted as the matching criterion for the subproblem-to-solution matching procedure.

S-CoEA_D_No: In this variant, the subproblem-to-solution matching procedure is removed. The distance value is still regarded as the matching criterion for the solution-to-subproblem matching procedure.

S-CoEA_D_D: In this variant, the distance value is adopted as the matching criterion for the subproblem-to-solution matching procedure for UOPs with uncertain constraints.

S-CoEA_CV_D: In this variant, the constraint violation value and the distance value are adopted as matching criteria for the solution-to-subproblem matching procedure and the subproblem-to-solution matching procedure, respectively.

S-CoEA_CV_CV: In this variant, the constraint violation value is adopted as the matching criterion for the solution-to-subproblem matching procedure.

All variants are the same as S-CoEA except for differences on two matching procedures. Due to limited space, only comparison results on the *Interval* test instances are illustrated. Taking the parameter settings as Section 4.3, the above six variants are experimentally compared with S-CoEA. Tables 4.3 to 4.4 summarize the overall performance of all comparison algorithms on three *Interval* test instances in terms of the Mann–Whitney U test and the Kolmogorov–Smirnov test, respectively. The overall performance includes the total number of solutions whose performance is worse than, better than, and similar to S-CoEA.

As can be seen from Tables 4.3 to 4.4, in terms of the Mann–Whitney U test, the proposed S-CoEA shows significant advantage over its variants on all *Interval* instances. In terms of the Kolmogorov–Smirnov test, similar results can be obtained except that their values are slightly different. Note

Table 4.3 Statistical Results of Various Variants Compared with S-CoEA over 30 Independent Runs on the *Interval* Test Instances [Jiang et al. (2008)] in Terms of the Mann–Whitney U Test.

Instance (†/§/≈*)	S-CoEA_No_No	S-CoEA_No_CV	S-CoEA_D_No	S-CoEA_D_D	S-CoEA_CV_D	S-CoEA_CV_CV
Interval1	**2178**/444/1008	**2979**/294/357	**3201**/72/357	912/618/**2100**	1629/258/**1743**	0/0/**3630**
Interval2	**2178**/1452/0	**1911**/1719/0	**2178**/1452/0	**2178**/1452/0	**2541**/1089/0	**2178**/1452/0
Interval3	**3888**/2592/0	**4383**/2097/0	**5184**/1296/0	**4536**/1944/0	**3944**/2536/0	**5184**/1296/0

Table 4.4 Statistical Results of Various Variants Compared with S-CoEA over 30 Independent Runs on the *Interval* Test Instances [Jiang et al. (2008)] in Terms of the Kolmogorov–Smirnov Test.

Instance (†/§/≈)	S-CoEA_No_No	S-CoEA_No_CV	S-CoEA_D_No	S-CoEA_D_D	S-CoEA_CV_D	S-CoEA_CV_CV
Interval1	**2681**/0/949	**2100**/0/1530	363/0/**3267**	363/0/**3267**	36/0/**3594**	528/291/**2811**
Interval2	**2178**/1452/0	**1911**/1719/0	**1815**/1452/363	**2178**/1452/0	**2541**/1089/0	**2178**/1452/0
Interval3	**3888**/2592/0	**4383**/2097/0	**5184**/1296/0	**4536**/1944/0	**3944**/2536/0	**5184**/1296/0

that for the *Interval1* test instance, each paired algorithms obtain different comparison results with respect to two performance measures, although both indicators are used to determine whether two independent samples are significantly different. Specifically, in terms of the Mann–Whitney U test, the majority of solutions obtained via S-CoEA are better than those of its variants. However, the majority of solutions obtained by S-CoEA show competitive performance compared with its variants in terms of the Kolmogorov–Smirnov test. The reason for this occurrence is that the two statistics hypothesis tests follow different assumptions.

In summary, the superiority of the proposed S-CoEA over its variants is highlighted on all *Interval* test instances in terms of two performance measures. Thus, the effectiveness of the solution-to-subproblem matching procedure using the distance value as the matching criterion and the subproblem-to-solution matching procedure adopting the constraint violation value as the matching criterion is confirmed experimentally.

4.5   Numerical Results on Continuous and Discrete Test Instances

This section is devoted to the experimental design for investigating the performance of S-CoEA on continuous and discrete test instances.

4.5.1   *Comparisons on Continuous Test Instances*

4.5.1.1   *Experimental results on EVM and Interval3*

There are no algorithms that deal with multiple uncertain models simultaneously in the literature. Therefore, for *EVM* and *Interval3* test instances, optimal solutions obtained in the original paper will be used for comparison against the results obtained via the proposed S-CoEA.[10]

Figure 4.3 illustrates the boxplots of the final solutions with the minimum objective function values within 30 runs found by S-CoEA and the method adopted in the original paper on *EVM* and *Interval3* test instances. For clarity, the results of 10 uncertain models obtained via S-CoEA are illustrated. Figure 4.3 shows that S-CoEA can find solutions whose performance are better than that of the solution obtained in the original for the *EVM*

---

[10]For a fair comparison for *EVM* and *Interval3* test instances, the number of function evaluations is set the same as the original. As to the remaining test instances, the choice of parameters is the same as Section 4.3.

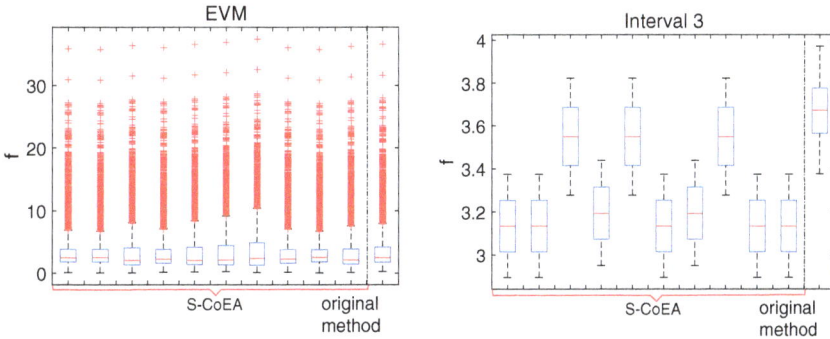

Fig. 4.3 Boxplots of the final solutions with the minimum objective function values within 30 runs found by S-CoEA and the method in the original paper on *EVM* and *Interval3* test instances.

test instance. For *Interval3*, S-CoEA shows obvious advantages over the comparison method.

4.5.1.2 *Experimental results on the remaining continuous test instances*

As to the remaining continuous test instances, there are only theoretical results in the literature. Two standard EAs with different evaluation strategies are adopted here as competitors. In the first evolutionary algorithm, denoted as EA1, the mean operator is adopted to transform uncertain functions into their deterministic counterparts. In the second single objective evolutionary algorithm, denoted as EA2, the worst-case operator is used. For both EA1 and EA2, the sample-updating scheme proposed in **Algorithm 2** is employed to gather landscape information of uncertain functions and alleviate the detrimental effects induced by uncertainties.

Here, the mean-case model and the worst-case model are selected to compare the performance of S-CoEA with EA1 and EA2, respectively. Similar to the experiments conducted in Section 4.3, uncertain parameters are randomly sampled 10,000 times for each test instance in order to calculate performance measures. Table 4.5 summarizes the overall performance of all comparison algorithms on the remaining continuous instances over 30 independent runs in terms of the Mann–Whitney U test and the Kolmogorov–Smirnov test.

Table 4.5 Statistical Results of Comparison Algorithms Compared with S-CoEA over 30 Independent Runs on the Remaining Continuous Instances in Terms of the Mann–Whitney U Test and the Kolmogorov–Smirnov Test.

Instance ($\dagger/\S/\approx$)		Mann–Whitney U Test		Kolmogorov–Smirnov Test	
		EA1	EA2	EA1	EA2
Test1	U	**30**/0/0	**30**/0/0	**30**/0/0	**30**/0/0
	N	**29**/1/0	**30**/0/0	**27**/3/0	**27**/3/0
Test2	U1	**30**/0/0	**30**/0/0	**30**/0/0	**28**/2/0
	U2	**30**/0/0	**29**/1/0	**30**/0/0	**25**/5/0
	U3	**24**/6/0	**30**/0/0	**29**/1/0	**30**/0/0
	N1	**30**/0/0	0/0/**30**	**30**/0/0	0/0/**30**
FNIM_f0		**30**/0/0	**30**/0/0	**26**/4/0	**30**/0/0
FNIM_f2	U1	**28**/2/0	**24**/6/0	**29**/1/0	**21**/9/0
	U2	**30**/0/0	**27**/3/0	**30**/0/0	18/0/**12**
	N1	**26**/4/0	10/8/**12**	**30**/0/0	9/0/**21**
	N2	**30**/0/0	18/**12**/0	**29**/1/0	18/**12**/0
FNIM_f3	U1	**30**/0/0	**27**/3/0	**30**/0/0	**24**/6/0
	U2	**30**/0/0	**30**/0/0	**30**/0/0	**30**/0/0
	N1	**30**/0/0	15/2/**13**	**30**/0/0	18/**12**/0
	N2	**30**/0/0	**30**/0/0	**30**/0/0	**30**/0/0
FNIM_f4	U1	**30**/0/0	**27**/3/0	**30**/0/0	**27**/3/0
	U2	**30**/0/0	**27**/3/0	**30**/0/0	**27**/3/0
	N1	**30**/0/0	**25**/0/5	**30**/0/0	**24**/6/0
	N2	**30**/0/0	**24**/6/0	**30**/0/0	**30**/0/0
Interval1		**30**/0/0	16/2/**12**	**30**/0/0	18/6/6
Interval2		**30**/0/0	14/5/**11**	**29**/1/0	3/0/**27**

As can be seen in Table 4.5, in terms of the Mann–Whitney U measure, the proposed S-CoEA shows significant advantages over EA1 on all test instances. S-CoEA performs clearly better than EA2 on the majority of test problems except for *Test2-N1*, on which S-CoEA and EA2 show similar performances. S-CoEA and EA2 perform competitively on nearly half of the uncertain models of *FNIM_f2-N1*, *FNIM_f3-N1*, *Interval1* and *Interval2*. Besides, S-CoEA and EA2 show competitive performance on *FNIM_f2-N2*. As to the Kolmogorov–Smirnov test, similar results can be obtained except that their values are slightly different on some instances. Note that for *FNIM_f2-U2*, the S-CoEA demonstrates similar performance on nearly half of the uncertain models as compared with EA2. As to *Interval2*, S-CoEA performs competitively on the majority of uncertain models as compared with EA2. To summarize, S-CoEA achieves significantly better performance than EA1 and EA2 in terms of the Mann–Whitney U test and the Kolmogorov–Smirnov test. The superiority of S-CoEA can be attributed to the efficient information sharing among neighboring subproblems and the two matching procedures that are beneficial to convergence.

For a visual observation, Fig. 4.4 shows the boxplots of the final solutions with the minimum objective function values within 30 runs found by S-CoEA, EA1 and EA2 on the remaining continuous test instances. Since the results about *Interval1* and *Interval2* are quite similar, only *Interval1* is selected and displayed. It is visually evident that for the majority of test

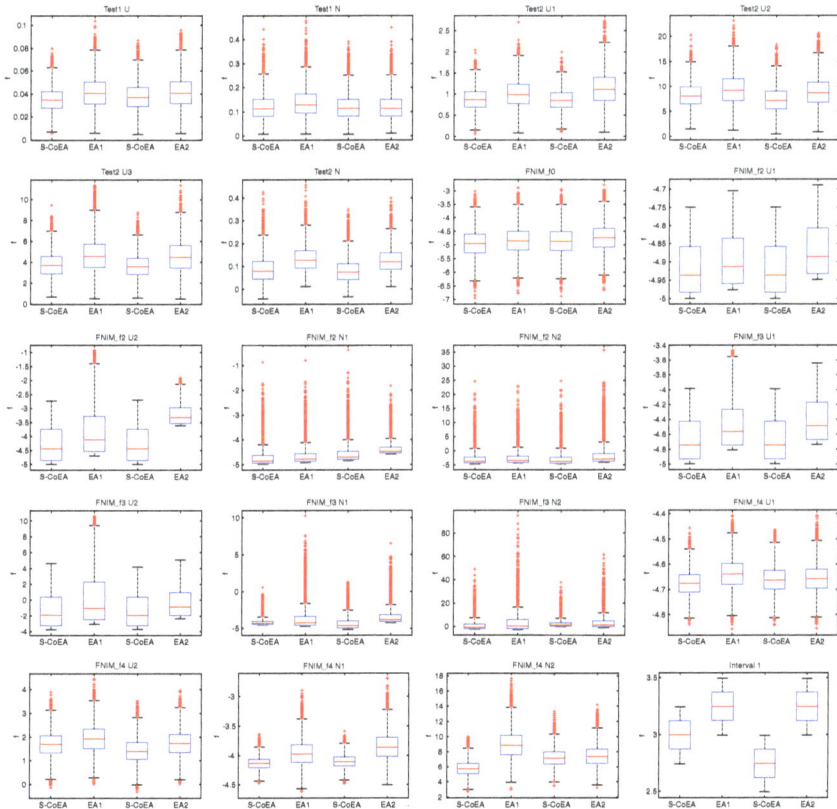

Fig. 4.4   Boxplots of the final solutions with the minimum objective function values within 30 runs found by S-CoEA, EA1 and EA2 on the remaining continuous test instances.

instances, S-CoEA can find solutions with better objective function values and smaller standard deviations as compared with EA1 and EA2.

4.5.2   *Comparisons on Discrete Test Instances*

Similar to Section 4.5.1, the mean-case model and the worst-case model are taken as an example to exhibit the comparison performance of S-CoEA with EA1 and EA2, respectively. In order to calculate two performance metrics, a set of 10,000 high density samples of uncertain parameters are randomly generated for each test instance. Table 4.6 summarizes the overall performance of all comparison algorithms on two suits of RAPs test instances over 30 independent runs in terms of the Mann–Whitney U test and the Kolmogorov–Smirnov test.

It is clear from Table 4.6 that S-CoEA performs significantly better than both EA1 and EA2 on *RAP1-U1* and *RAP1-U2* in terms of both the Mann–Whitney U test and the Kolmogorov–Smirnov test. As to *RAP2-U1*, S-CoEA outperforms EA1 and EA2 on the majority of uncertain models in terms of the Mann–Whitney U test. In terms of the Kolmogorov–Smirnov test, S-CoEA shows competitive performance in comparison with EA1, but performs better than EA2 on the majority of uncertain models. For *RAP2-U2*, S-CoEA demonstrates clear superiority only on half of the uncertain models, while the compared algorithm performs better than S-CoEA on the other half of uncertain models in terms of the Mann–Whitney U test. S-CoEA performs competitively against EA1 but shows clear advantages over EA2 on *RAP2-U2* according to the Kolmogorov–Smirnov test.

Figure 4.5 displays the boxplots of the final solutions with the minimum objective function values within 30 runs found by S-CoEA, EA1 and EA2 on RAPs test instances. From these figures, it is also clear that S-CoEA,

Table 4.6   Statistical Results of Comparison Algorithms Compared with S-CoEA over 30 Independent Runs on RAPs Instances in Terms of the Mann–Whitney U Test and the Kolmogorov–Smirnov Test.

Instance ($\dagger/\S/\approx^*$)		Mann–Whitney U Test		Kolmogorov–Smirnov Test	
		EA1	EA2	EA1	EA2
RAP1	*U1*	**21**/9/0	**21**/9/0	**30**/0/0	**25**/5/0
	U2	**24**/6/0	**30**/0/0	**27**/1/0	18/6/6
RAP2	*U1*	**24**/6/0	**23**/4/3	18/12/0	**24**/5/1
	U2	18/12/0	13/11/6	**17**/12/1	**24**/6/0

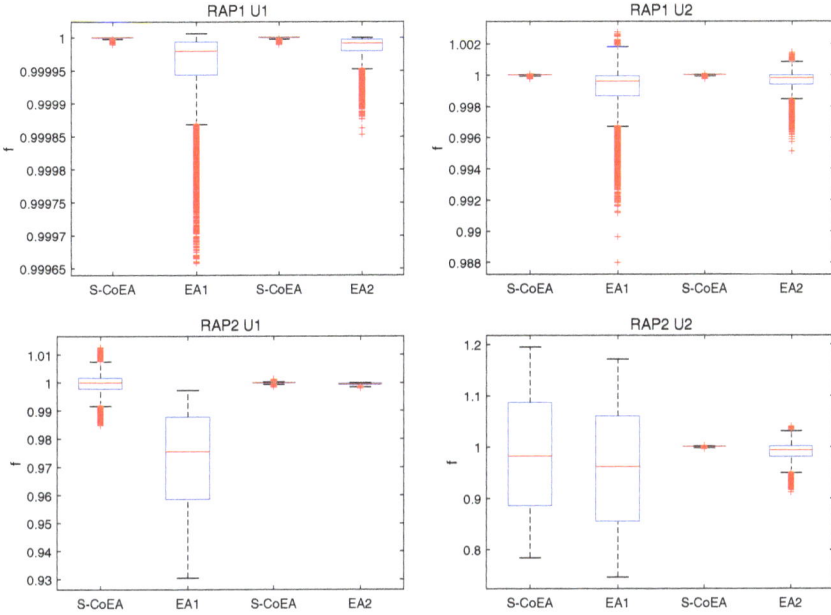

Fig. 4.5   Boxplots of the final solutions with the minimum objective function values within 30 runs found by S-CoEA, EA1 and EA2 on RAPs test instances.

EA1 and EA2 can get similar objective function values. However, solutions obtained by S-CoEA have smaller variance on the majority of test instances, which is preferable in reality.

4.6   Further Discussion

In this section, the parameter analysis of S-CoEA, including the influence of parameters T and I, is deeply analyzed.

4.6.1   *Parameter Sensitivity Analysis about T*

T represents the neighborhood size and is an important parameter in the S-CoEA. To study how S-CoEA is sensitive to this parameter, we take *Test1-U*, *FNIM_f2-N1*, *FNIM_f4-U1* and *Interval1* as examples and test different settings of T in the implementation of S-CoEA. Different T values are set as $\lfloor (10\%, 20\%, 40\%, 60\%, 80\%, 90\%, 100\%) \cdot N \rfloor$. All the other parameters are kept the same as Section 4.3. Similarly, 30 independent runs have been conducted for each configuration on these test instances. Here, only the mean models of selected test instances are chosen for illustration.

Figure 4.6 shows the variation of the means and variances of objective values across all T values on the selected test problems. As shown in Fig. 4.6, S-CoEA performs differently with different T values on *FNIM_f2-N1*, *FNIM_f4-U1* and *Interval1*, and it performs well with a wide range of T value on *Test1-U*. For all selected test instances, a large T value is better. Thus, it can be claimed that $\lfloor 90\% \cdot N \rfloor$ is a good choice for most of the test instances. Generally, a larger value of T is good for convergence, while a smaller value of T benefits the diversity of the evolving population.

4.6.2   *Parameter Sensitivity Analysis about I*

I is a major parameter in S-CoEA. It decides how often the algorithm updates the samples of uncertain parameters. This part studies how S-CoEA is sensitive to this parameter. We still take *Test1-U*, *FNIM_f2-N1*, *FNIM_f4-U1* and *Interval1* as examples and test different settings of I in the implementation of S-CoEA. Different I values are set to $\lfloor (1\%, 2\%, 5\%, 10\%, 20\%, 40\%, 60\%, 80\%, 90\%, 100\%) \cdot NFE \rfloor$. All the other parameters are kept the same as Section 4.3. Similarly, 30 independent runs have been conducted for each configuration on these test instances. The mean models of selected test instances are chosen for illustration.

Figure 4.7 shows the variation of the means and variances of objective values across all I values on the selected test problems. As can be seen from Fig. 4.7, S-CoEA performs differently with different I values on all selected test instances, and it performs well with a small I on all selected test instances. Generally, a large I value may mislead search processes because of a lack of information on uncertain functions. However, a small I value means more times in updating samples and performing the solution-to-subproblem matching procedure. This will result in more consumption of computational resources. Thus, a proper I value strikes a good balance between the performance of S-CoEA and its computational cost.

4.7   Conclusion

To our best knowledge, current existing techniques on dealing with UOPs are all *a priori* methods. Actually, it may be obtrusive or even risky to incorporate the preferences of a DM to handle uncertainties when he/she does not have sufficient knowledge about the problem. This chapter reformulates UOPs and proposes an *a posteriori* decision-making framework

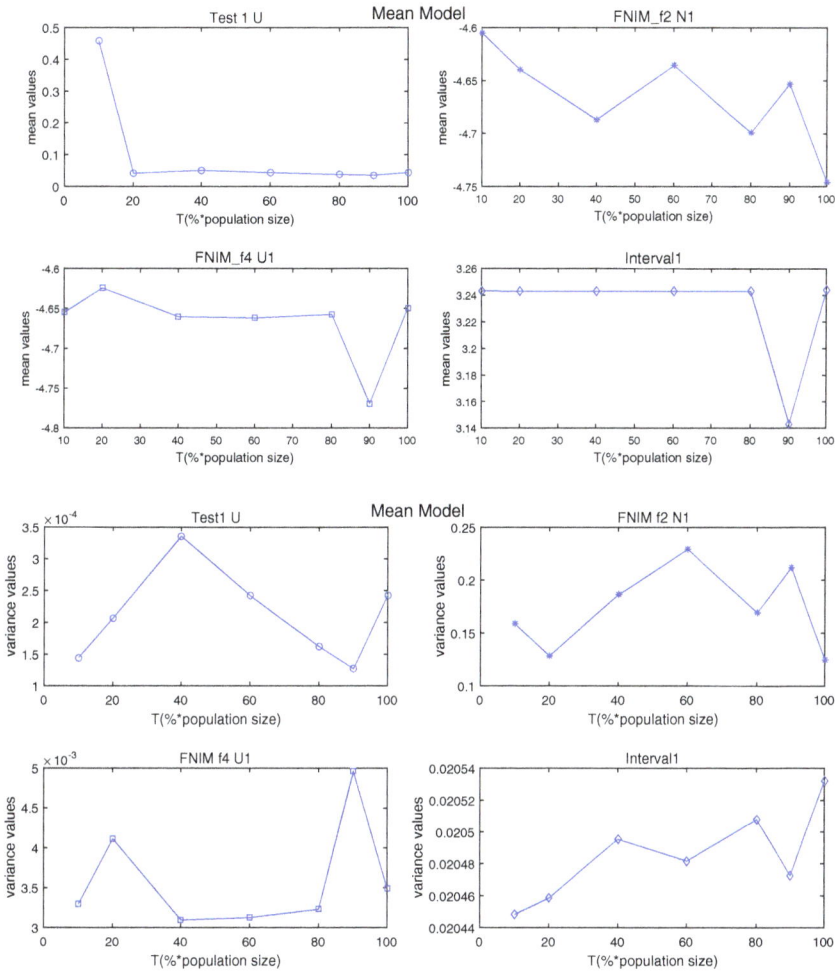

Fig. 4.6   The means and variances of objective values within 30 runs versus the value of T in S-CoEA for *Test1-U*, *FNIM_f2-N1*, *FNIM_f4-U1*, *Interval1* and *Interval3* test instances.

that covers several common uncertain models without using preferences on handling uncertainties for UOPs. Furthermore, we incorporate the weighting method into a decomposition strategy and propose a subproblems co-solving evolutionary algorithm, i.e., S-CoEA to cope with UOPs. S-CoEA explicitly decomposes a UOP into a series of subproblems that represent

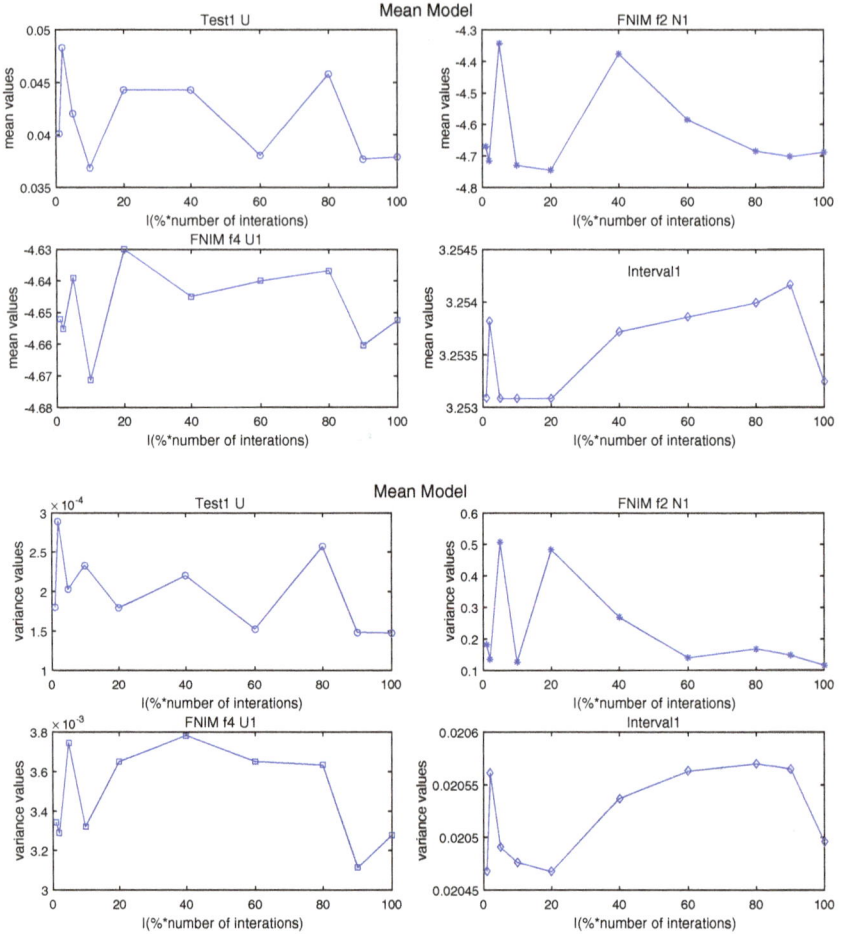

Fig. 4.7   The means and variances of objective values within 30 runs versus the value of *I* in the S-CoEA for *Test1-U*, *FNIM_f2-N1*, *FNIM_f4-U1*, *Interval1* and *Interval3* test instances.

different uncertain models by assigning each subproblem with a weight vector. Then, these subproblems are solved in parallel by evolving a population of solutions. At each generation, each individual solution in the population is associated with a subproblem. The neighborhood relations among these subproblems are defined based on the Euclidean distance between their weight vectors. Besides, a sample-updating strategy based on historical information is proposed for UOPs. A solution-to-subproblem matching procedure and a subproblem-to-solution matching procedure are proposed to enhance convergence.

S-CoEA has been compared with its various variants on continuous test instances to demonstrate the superiority of two matching procedures. Besides, S-CoEA has also been compared with competitors on continuous test instances and RAPs with various characteristics and different strength levels of uncertainties. A systematical experimental study has shown that S-CoEA outperforms or performs competitively against comparison algorithms on the majority of test instances. Moreover, the parameter sensitivity analysis about T and I in S-CoEA has been experimentally investigated. All these experimental results confirm that S-CoEA can deal with the majority of continuous and discrete benchmark problems.

Future research work includes investigations of trying other methods of updating samples, hybridizing other mechanisms of handling uncertainties in S-CoEA to further improve its performance, and extending it to uncertain MOPs.

Chapter 5

Noise-Tolerant Techniques for Decomposition-based Multi-objective Evolutionary Algorithms

Multi-objective evolutionary algorithms (MOEAs) are mainstream methods for tackling multi-objective problems (MOPs). Among various MOEAs, decomposition-based MOEAs (DMOEAs) are growing in popularity and have became major methodologies for approximating Pareto fronts (PFs), thanks to the success of the MOEA/D framework. However, the presence of noise in objective functions may affect the ability of algorithms to drive the search process toward the PF, since it is difficult to definitely determine the Pareto dominance relationship between two solutions. Thus it is necessary to design noise-tolerant MOEAs that can perform their actions notwithstanding the presence of noise. Evolutionary algorithms (EAs) are known to be inherently robust to low-level noise due to their distributed nature and nonreliance on gradient information. However, such a property may not extend well into MOEAs, since MOEAs require the evolutionary search process to maintain a set of non-dominated solutions uniformly distributed along the PF [Goh and Tan (2007)]. There are quite a few studies on designing algorithms for UMOPs [Boonma and Suzuki (2009); Buche *et al.* (2002); Fieldsend and Everson (2005, 2015); Goh and Tan (2007); Knowles *et al.* (2009); Rakshit *et al.* (2014); Siwik and Natanek (2008)]. Fieldsend and Everson (2005) put forward the Bayesian $(1 + 1)$-ES (BES) that assesses the probability of dominance and maintains a set of mutually non-dominated solutions with a predefined probability. Goh and Tan (2007) proposed an MOEA with three robust features that include an experimental learning directed perturbation, a gene adaptation selection strategy, and a possibilistic archiving model based on the concept of possibility and necessity measures. Besides, a novel algorithm, named the rolling tide evolutionary algorithm (RTEA), was developed by Fieldsend and Everson (2015). The RTEA progressively improves the accuracy of its estimated Pareto set, while simultaneously driving the population towards the true PF.

The most common strategy for noisy problems is sampling. There are many studies on dynamic adjusting sample size for single objective problems (SOPs) [Branke and Schmidt (2003); Cantú-Paz (2004)]. However, the extension of these methods to MOPs is not straightforward. Only a few number of studies focus on the issue of adaptive sample size for MOPs [Park and Ryu (2011); Rakshit *et al.* (2014); Syberfeldt *et al.* (2010)]. For example, in [Rakshit *et al.* (2014)], a linear relationship between the sample size of a trial solution and the fitness variance in its local neighborhood was employed for non-uniform sampling. Syberfeldt *et al.* (2010) proposed a confidence-based dynamic sampling technique that varies the number of samples used per solution based on the amount of noise in combination with a user-defined confidence level. Most methods adopt the mean value of fitness samples as the fitness estimation of a trial solution to improve the accuracy of estimation. Rakshit and Konar (2015a,b) proposed that the expected value of fitness samples be determined on the basis of their distribution as the fitness measure of a trial solution.

Other noise-handling techniques in optimizing UMOPs include periodic re-evaluation of archived solutions [Buche *et al.* (2002)], probabilistic Pareto ranking [Hughes (2001)], possibilistic archiving [Goh and Tan (2007)], fitness inheritance [Bui *et al.* (2005)], extended averaging scheme [Singh (2003)], and so on. Buche *et al.* (2002) proposed to modify the elite preservation scheme with the aim of reducing detrimental effects of outliers for noisy combustion processes. In particular, every solution is assigned a lifetime that is dependent on the fraction of the archive it dominates. Any archive solutions with expiring lifetime are re-evaluated and added to the evolving population. In the subsequent archive updating procedure, expired solutions will not be considered. Hughes (2001) demonstrated the possible deficiencies of the non-dominated sorting approach and then introduced a probabilistic Pareto ranking scheme to account for noisy and uncertain systems. Singh (2003) proposed an extended averaging scheme to reduce the bias introduced by the small sample size in the optimization of groundwater remediation design. The extended averaging approach performs the averaging over all samples of identical individuals, which can be easily extended over different generations.

This chapter first examines the performance of DMOEAs in noisy environments. It has been observed that the impact of noise on DMOEAs are different for benchmarks with different characteristics and different strength levels of noise. That is, DMOEAs tend to perform well for the majority of problems in the presence of low-level noise, and the evolutionary optimization process degenerates into a random search under increasing level of noise. While for some problems, the evolutionary optimization process is badly effected even by the presence of low-level noise. Based on the analyses of noise impacts on population dynamics of convergence and diversity, four noise-handling techniques including a Pareto-based nadir point estimation strategy, two adaptive sampling strategies, a mixed objective evaluation strategy, and a mixed repair mechanism are proposed and incorporated into two existing DMOEAs to improve their performances in the presence of noise. The differences between the proposed approach and existing methodologies are the way they deal with uncertainties.

5.1   Impacts of Noise on DMOEAs

In this subsection, a brief description of benchmark problems, performance metrics, and information on experimental setup are provided at first. Then we examine impacts of noise on the dynamic of convergence and diversity over generations in DMOEAs and present the motivations of this chapter.

5.1.1   *Benchmark Problems*

Benchmark problems are used to reveal the capabilities and important characteristics of algorithms under evaluation. In the context of MOPs, researchers have identified several benchmarks with various characteristics. Among them, 5 ZDT test instances [Zitzler *et al.* (2000)], 2 tri-objective DTLZ test instances [Deb *et al.* (2002b)], and 10 UF test suites [Zhang *et al.* (2009b)], including ZDT1-ZDT4, ZDT6, DTLZ2, DTLZ4 and UF1-UF10 are adopted in this chapter. These test instances are used to examine the effectiveness of DMOEAs in converging and maintaining a diverse set of non-dominated solutions under the influence of noise. In this study, noise is implemented as an additive perturbation on the objective value of each individual [Hughes (2001)], i.e., $\tilde{f}_i(\mathbf{x}, \alpha, \eta) = f_i(\mathbf{x}, \alpha) + \eta_i$, $i = 1, \ldots, m$, where m stands for the number of objectives. η_i is the additive noise of the

ith objective function, which is often assumed to be normally distributed with a zero mean and variance σ_i^2.[1] σ_i^2 means the strength level of noise present in the ith objective function and is often represented as a percentage of $|F_i^{\max} - F_i^{\min}|$, where $F_i^{\max}$ and $F_i^{\min}$ are the maximum and minimum of the ith objective function in the true PF [Goh and Tan (2007)]. $\tilde{f}$ and f denote the objective function with and without additive noise, respectively. Test functions of MOPs will be modified in the above form in order to include the influence of noise.

Ideally, MOEAs should work on the expected fitness function $E[\tilde{f}_i(\mathbf{x}, \alpha, \eta)]$ and not be misled by the presence of noise. During optimization, the only measurable fitness value is a stochastic value $f_i(\mathbf{x}, \alpha) + \eta_i$. Therefore, in practice, the expected fitness value $E[\tilde{f}_i(\mathbf{x}, \alpha, \eta)]$ is often approximated by the averaged sum of a number of random samples: $\hat{f}_i(\mathbf{x}, \eta) = \frac{1}{n} \sum_{i=1}^{n} (f_i(\mathbf{x}) + \eta_{i,j})$, $i = 1, \ldots, m$, where $\{\eta_{i,1}, \ldots, \eta_{i,n}\}$ are n samples of the additive noise of the ith objective function, and $\hat{f}$ is an unbiased estimation of f.

5.1.2   *Performance Metrics*

Performance metrics pertinent to the convergence and diversity play an important role when evaluating the quality of a set of obtained non-dominated solutions for MOPs. According to the law of large numbers, the mean value of sampled objective values converges to the noise-free objective function value when the number of samples becomes large enough for any noisy function. Thus true PFs of noisy and deterministic MOPs are the same, and thus performance metrics used in deterministic MOPs are still valid in noisy cases. Here two performance metrics, i.e., the averaged Hausdorff distance (Δ_p) [Schutze *et al.* (2012)] and the hypervolume (HV) [Zitzler and Thiele (1999)] are employed.

5.1.3   *Parameter Settings*

Both MOEA/D and DMOEA-εC are adopted in this section to examine impacts of noise on DMOEAs. They employ a fixed-size population and

---

[1]Additional numerical experiments on UMOPs with uniform noise are conducted to test the performances of comparison algorithms on non-Gaussian noise. Numerical results have concluded that no qualitative difference has been observed in the presence of Gaussian and non-Gaussian noise, which is consistent with the statement in [Jin and Branke (2005)].

an archive that stores non-dominated solutions along the evolution. The archive is updated at each cycle, i.e., a candidate solution will be added into the archive if it is not dominated by any member in the archive. Likewise, any archive member dominated by this solution will be removed from the archive. When the predetermined archive size is reached, a truncation process based on the crowding distance [Deb *et al.* (2002a)] is used to eliminate the most crowded archive member. Besides, the dynamic resource allocation strategy [Zhang *et al.* (2009a)] is added into both MOEA/D and DMOEA-εC. Other implementations adopted here of the two algorithms are the same as those in [Chen *et al.* (2017)].

Experiments are conducted at noise strength levels of $\sigma^2 \in \{0.1\%, 0.2\%, 0.5\%, 1\%, 5\%, 10\%, 20\%\}$ in order to study impacts of noise on DMOEAs. For a fair comparison, the choice of parameters are the same for both MOEA/D and DMOEA-εC. Specifically, the population size is set to $N = 100$ for ZDT problems. As to the DTLZ problems, due to the differences in algorithmic frameworks, N is set to 351 and 324 for MOEA/D and DMOEA-εC, respectively. For UF instances, the population size N is set to 600 and 1000 for bi-objective and tri-objective instances, respectively. The size of the archive is set as $S = N$. Besides, the DE operator and Gaussian mutation are used in solving ZDT and DTLZ test problems, and the DE operator and polynomial mutation are adopted for UF instances [Chen *et al.* (2017)]. In order to eliminate effects of noise, the uniform sampling strategy is adopted, and the number of samples is set $n = 5$ evenly for each subproblem at each generation. Both algorithms terminate when the number of evaluations reaches the maximum number. Based on the parameter settings in deterministic cases, the maximum number of function evaluations are set as 250,000, 375,000 and 1500,000 for ZDT, DTLZ and UF instances, respectively [Chen *et al.* (2017)]. Finally, each algorithm is executed 30 times independently on each instance.

Two performance metrics, i.e., Δ_p and HV are employed to evaluate the performance of all comparison algorithms. With the purpose of calculating the Δ_p metric value, P^* is chosen to be a set of 500 uniformly distributed points along the true PF for ZDT problems, and 1024 points for DTLZ instances. As to the bi-objective UF, a set of 1000 uniformly distributed points along the true PF are chosen as P^*, except that 21 uniformly distributed points are chosen as P^* for UF5. For tri-objective UF test problems, P^* is chosen to be a set of 10,000 uniformly distributed points along the true PF [Chen *et al.* (2017)]. Besides, in order to compute the HV metric value, the reference point is set as 1.1 times the true nadir point.

5.1.4   *Empirical Results of Noise Impacts on DMOEAs*

Figures 5.1 and 5.2 illustrate distributions of the final populations in the objective space with the minimum Δ_p metric value within 30 runs found by MOEA/D and DMOEA-εC on 4 test problems under the influence of 0.5%, 5% and 20% noise levels, respectively. Both MOEA/D and DMOEA-εC obtain solutions uniformly spread along the true PF on the ZDT2 and UF2 test instances with a 0.5% noise level except UF2, on which MOEA/D can only cover a part of the true PF. As to the tri-objective DTLZ2 and UF10 test instances with a 0.5% noise level, DMOEA-εC performs better than MOEA/D and obtains final populations that will cover the whole PF but with bad uniformity. According to the observation of Nissen and Propach (1998) population-based EAs are inherently robust in single objective optimization under a low-level noise. It is also confirmed from Figures 5.1 and 5.2 that DMOEAs are capable of obtaining satisfactory solutions on some test instances under the influence of low-level noise. When it comes to the high-level noise, i.e., 5% and 20% noise levels, both MOEA/D and

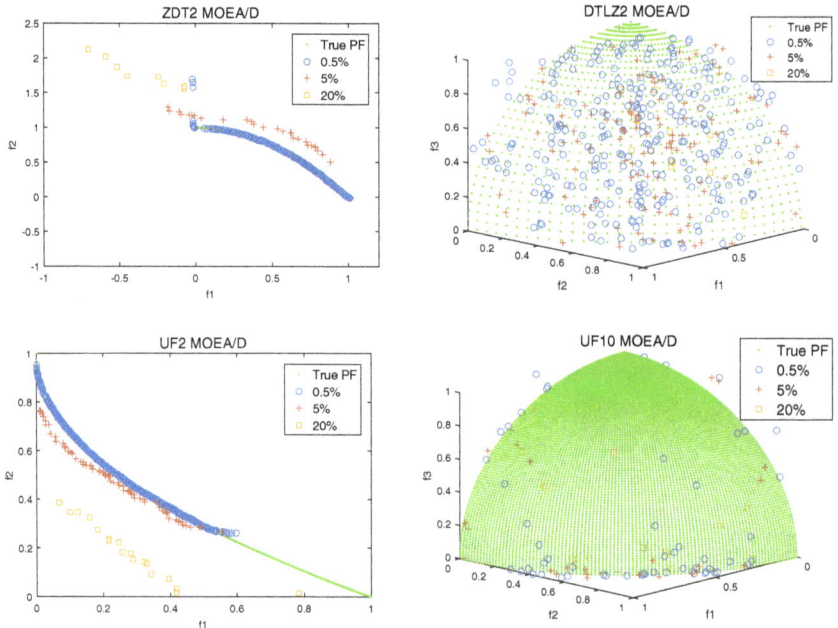

Fig. 5.1   Final populations in the objective space with the minimum Δ_p metric value within 30 runs obtained by MOEA/D on 4 test problems under the influence of 0.5%, 5%, and 20% noise levels.

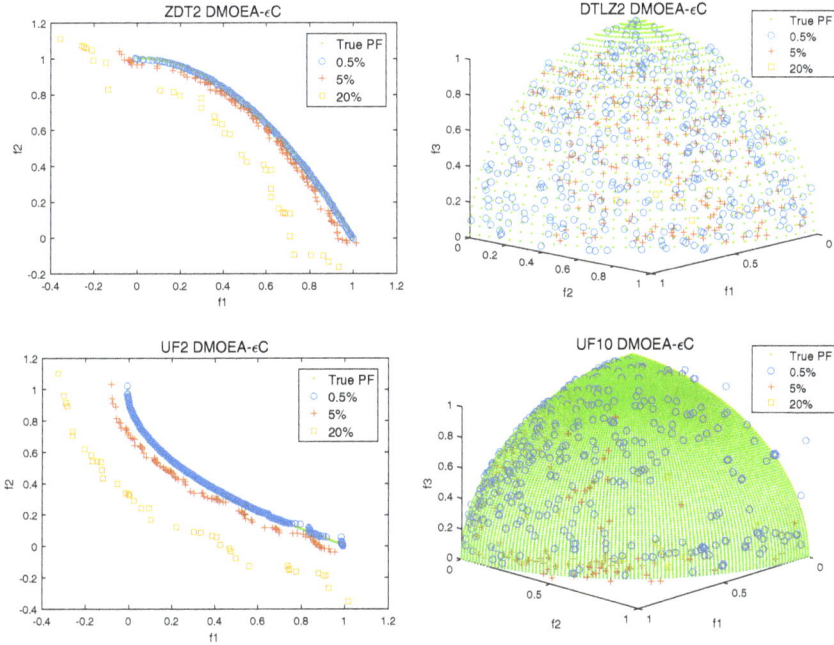

Fig. 5.2 Final populations in the objective space with the minimum Δ_p metric value within 30 runs obtained by DMOEA-εC on 4 test problems under the influence of 0.5%, 5% and 20% noise levels.

DMOEA-εC cannot achieve approximations with both good convergence and diversity on the ZDT2 and UF2 test instances. Besides, for DTLZ2 and UF10 with high-level noise, final solutions obtained via MOEA/D and DMOEA-εC can hardly approximate and cover the whole PF very well.

In order to further examine algorithmic behaviors in the objective space, objective values during the search process are recorded. The evolution of objective values with the number of iterations obtained via DMOEA-εC on DTLZ2 with 0.1% and 20% noise levels are presented in Fig. 5.3. The traces of objective values are sufficient to demonstrate impacts of noise on multiple objective values during the evolutionary process. It can be seen from Fig. 5.3 that the introduction of high-level noise degrades the performance of DMOEA-εC on DTLZ2. Besides, it should be noted that DMOEA-εC stagnates in the later stage of evolution on DTLZ2 with 0.1% and 0.2% noise levels. This fact has been stated in [Goh and Tan (2007)] that basic EAs suffer from degenerate convergence properties and face

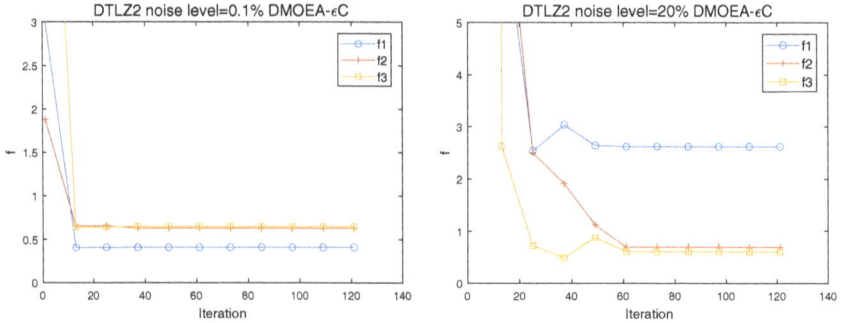

Fig. 5.3   Traces of three objective values obtained via DMOEA-εC for DTLZ2 under the influence of 0.1% and 20% noise levels.

the problem of maintaining a diverse solution set under the influence of noise.

Generally, when using the sampling strategy to deal with noisy problems, the mean value over several samples is regarded as the fitness estimation to eliminate detrimental effects induced by noise. However, other statistics can also be adopted as criteria when evaluating a solution in noisy environments. We conduct experiments on all test instances to see the differences of final populations obtained via DMOEAs with the mean criterion and with the median criterion. Due to the lack of space, results obtained via MOEA/D on three instances with different noise levels are illustrated in Fig. 5.4. It shows that MOEA/D with mean and median criteria perform similarly when the noise level is low. When the noise level increases, MOEA/D with the mean criterion tends to obtain solutions with good convergence. However, solutions found by MOEA/D with the median criterion show good diversity. Based on these observations, it can be concluded that the mean and median criteria have their advantages when tackling noisy problems. It is natural to have an idea of combining them together and obtaining a mixed criterion that performs well on all noise levels.

As stated above, DMOEAs stagnate at the later evolution stage because of diversity loss. Reproduction operators are the primary source of new solutions and should be responsible for the loss of diversity in the decision space, which will lead to the loss of diversity in the objective space. We notice that the truncation repair mechanism is commonly used when solutions exceed search boundaries. Truncation repairing means that each element of a solution that exceeds the search space will be truncated back to the edge of

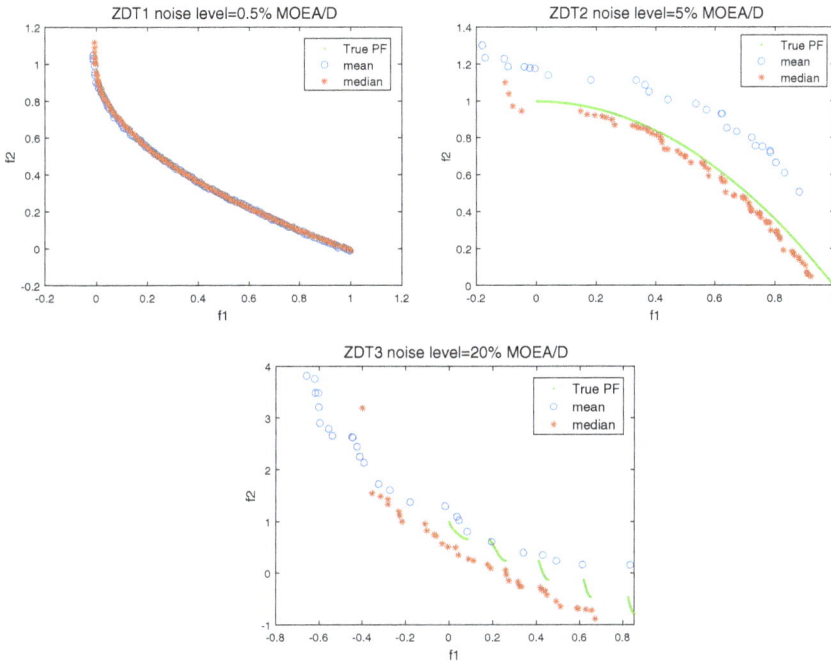

Fig. 5.4   Final populations in the objective space with the minimum Δ_p metric value within 30 runs obtained by MOEA/D using the mean and median values as objectives estimation on 3 instances under the influence of different noise levels.

its domain. Such a fixed form of repair mechanism is not conducive to the diversity in the decision space, and the effect of losing diversity is more outstanding when handling noisy problems. Random repair methods mean each element of solution that exceeds the search space will be replaced with a random feasible value. We try to replace the truncation repair method with a random one and test the performances of DMOEAs with the random repair mechanism on all test problems. Detailed results obtained via DMOEAs with the random repair mechanism on all test instances with different noise levels are omitted due to the limited space. It is concluded that usage of the random repair mechanism will enhance the diversity of decision space, which will also lead to more diverse final populations. Based on these observations, it is natural to have an idea of combining the two repair mechanisms together and proposing a mixed repair mechanism that performs well at all levels of noise. The rationality and superiority of the mixed criteria and the mixed repair mechanism will be shown in Section 5.3.2.

5.2 Noise-tolerant DMOEAs

From observations in the previous section, it is confirmed that the performance of DMOEAs deteriorates sharply at high level noise. This section extends and improves standard DMOEAs by four noise-handling features based upon the analyses of population dynamics under the influence of noise.

5.2.1 *Pareto-based Nadir Point Estimation Strategy*

Ideal points and nadir points have important impacts on determining the search range in the objective space of DMOEAs. Imprecise estimations may mislead the search direction of the evolutionary process and result in approximations of only parts of PFs. Nadir points are much harder to obtain than ideal points. Several approaches have been proposed to calculate nadir points [Alves and Costa (2009); Metev and Vassilev (2003)]. In standard DMOEAs, ideal points and nadir points are replaced by the minimum value of each objective function in the population and the maximum value of each objective function in the external archive, respectively. Besides they will be updated iteratively. However, this procedure is not effective enough in noisy environments.

Pareto-based and decomposition-based MOEAs are two prevailing approaches for MOPs that have their own advantages and disadvantages [Li *et al.* (2015b)]. Specifically, Pareto-based MOEAs are featured with fast convergence at the early stage of optimization. Along with the searching process of Pareto-based MOEAs, the majority of solutions in evolving population become non-dominated and the concept of Pareto dominance cannot distinguish these solutions. Then diversity maintaining mechanisms, instead of Pareto dominance relationships, become the main power that pushes an evolving population towards the true PF, which slows down the convergence at the late stage of optimization. DMOEAs do not suffer from the ineffectiveness of Pareto dominance relationship but need information of ideal and nadir points [Chen *et al.* (2017); Wang *et al.* (2017)]. Thus it is reasonable to obtain an estimation of the nadir point by taking the advantage of Pareto-based MOEAs at the early stage of optimization. With the guidance of a good estimated nadir point, DMOEAs are adopted to evolve the population towards the true PF with a quick convergence speed. By combining strengths of two MOEAs, not only a good estimation of the nadir point, but also a well initialized population can be obtained. Here NSGA-II is selected as a representative of Pareto-based MOEAs.

5.2.2 *Adaptive Sampling Strategies*

One common approach for uncertain problems is to perform sampling for individuals, otherwise the evolutionary selection process may become unstable. A crucial aspect here is to find the best trade-off between the number of solutions evaluated and the number of samplings for each solution. The larger the number of solutions evaluated, the more the search space can be explored and the greater the probability of finding its optimum will be. However, at the same time, sampling of solutions is necessary in order to prevent the search from being misdirected by noise. It is not wise to equally sample all candidate solutions with different qualities, especially when a small number of samplings is allowed. Many researches on adaptively adjusting the sample size of each solution have been done on uncertain SOPs. These techniques cannot be directly extended to UMOPs since expected solutions of UMOPs are a set of solutions with good convergence and diversity. However, the development of DMOEAs bridges SOPs with MOPs, and thus some sample size adjustment strategies designed for uncertain SOPs can be extended to UMOPs. Here, we introduce two typical adaptive sampling strategies, i.e., the co-ranking allocation strategy [Cantú-Paz (2004)] and the optimal computing budget allocation (OCBA) strategy [Chen (1995); Chen *et al.* (1999)].

In DMOEAs, an MOP is converted into a series of scalar subproblems, and each subproblem is associated with a solution in the population and a set of neighboring solutions that can be regarded as the population of each subproblem. The neighborhood relations among these subproblems are defined based on the Euclidean distance between their weight vectors or upper bound vectors ε^i ($i = 1, \ldots, N$), where N is the number of subproblems. The neighborhood of the ith subproblem consists of all subproblems with the weight vectors or upper bound vectors from the neighborhood of ε^i. In the proposed noise-tolerant DMOEAs, the concept of neighborhood is used for two main purposes including the update of neighboring solutions and the update of samples. Thus we define two types of neighborhoods for each subproblem. The neighborhoods of the ith subproblem for updating solutions and updating samples are denoted as $B_1(i)$ and $B_2(i)$, respectively.

In both adaptive sampling strategies, each subproblem has a predefined sample size and samples will be allocated among its neighboring solutions. It should be noted that since a subproblem falls within neighbourhoods of several subproblems, the final sample size of each subproblem is determined after adaptive sampling strategies are done for all subproblems. Then the estimated objective value of a solution is given based on the adaptive sample size and the mixed objective evaluation strategy, which will be presented below.

5.2.2.1   *Co-ranking adaptive sampling strategy*

In order to ensure a high probability to find the best solution, a larger sample size is adopted for individuals with a higher estimated variance. Besides, with the purpose of distinguishing two solutions, a larger number of samplings should be conducted on two solutions with similar performance. The difference between two solutions is measured with the absolute value of the difference between two estimated objective values, as described in line 7 of **Algorithm 13**. The estimated objective value of a solution can be the mean, median or mixed criterion value over a number of sampled objective values. Thus a synthetical evaluation of each solution is conducted by a weighted aggregation of rank values of the variance and the difference value between neighboring solutions. Different weighted coefficients α mean different aggregation forms. For example, $\alpha = 1$ means allocating a sample size that completely depends on the variance value of each solution. However, $\alpha = 0$ stands for the adaptive sample size that is only related to the difference between two solutions. Here, we emphasize two factors equally and adopt $\alpha = 0.5$.

5.2.2.2   *OCBA strategy*

The OCBA technique proposed by Chen and colleagues [Chen (1995); Chen *et al.* (1999)] is used to optimally choose the number of simulations for all of the solutions in order to maximize simulation efficiency with a given computing budget. Here the OCBA technique is applied to UMOPs to allocate limited sampling budgets among solutions to provide reliable evaluation and identification of good individuals. Details of the OCBA sampling strategy are illustrated in **Algorithm 14**. Similar to the co-ranking sampling strategy, the variance value and the difference value are two main components in the OCBA strategy. However, the OCBA strategy uses them in a different way and does not need an extra coefficient α. Readers can refer to [Chen (1995); Chen *et al.* (1999)] for details of the OCBA technique.

5.2.3   Remedy for the Loss of Diversity

It has been observed in Section 5.1.4 that DMOEAs suffer from degenerate convergence properties and face the problem of maintaining diversity of

---

Algorithm 13 Co-ranking allocation strategy

Require: The predefined sample size for each subproblem k, a weighting coefficient α, and N subproblems.

Ensure: Sample size for N subproblems, i.e., $q_1, \ldots, q_N$.

1: Initialize $q_i = 0, i = 1, \ldots, N$.

2: **for** $i = 1$ to N **do**

3:     Get the neighborhood of updating samples for the ith subproblem, i.e., $B_2(i)$.

4:     **while** $B_2(i)$ is nonempty **do**

5:        Randomly select an index $j \in B_2(i)$.

6:        Calculate the variance of jth subproblem var_j.

7:        Calculate the difference between the two subproblems $dis_j = |\bar{g}^i - \bar{g}^j|$, where $\bar{g}^i$ and $\bar{g}^j$ are estimated objective values of the ith and jth subproblems.

8:        $B_2(i) = B_2(i) \backslash \{j\}$.

9:     **end while**

10:     Sort $\{var_1, \ldots, var_{|B_2(i)|}\}$ in a descending order and obtain the corresponding rank value of each subproblem $rank_j^v, j \in B_2(i)$.

11:     Sort $\{dis_1, \ldots, dis_{|B_2(i)|}\}$ in a descending order and obtain the corresponding rank value of each subproblem $rank_j^d, j \in B_2(i)$.

12:     **for** $j = 1$ to $|B_2(i)|$ **do**

13:        $rank_j = \alpha \cdot rank_j^v + (1 - \alpha) \cdot rank_j^d$. // $rank_j$ is the overall rank value of the jth subproblem.

14:     **end for**

15:     Initialize $count = 0$.

16:     **while** $count \leq k$ **do**

17:        $s = \operatorname{argmin}_{l=1,\ldots,|B_2(i)|}(rank_1, \ldots, rank_{|B_2(i)|})$.

18:        $q_s = q_s + 1; count = count + 1$.

19:        $B_2(i) = B_2(i) \backslash \{s\}$.

20:     **end while**

21: **end for**

---

population under the influence of noise. Besides, Section 5.1.4 also reveals that the usage of the median value as the objective estimation and the random repair mechanism both can improve the diversity of an evolving population. Since DMOEAs show similar performances on test instances with a noise level smaller than 1%, between 1% and 10%, and larger than or equal to 10%, seven levels of noise are classified into three categories: low-, medium-, and high-level noise, which is rational in practice.

Algorithm 14 OCBA Strategy

Require: The sample size for each subproblem k and N subproblems.
Ensure: Sample size for N subproblems, i.e., $q_1, \ldots, q_N$.

1: Initialize $q_i = 0, i = 1, \ldots, N$.
2: **for** $i = 1$ to N **do**
3:      Get the neighborhood of updating samples for the ith subproblem, i.e., $B_2(i)$.
4:      **while** $B_2(i)$ is nonempty **do**
5:          Randomly select an index $j \in B_2(i)$.
6:          Calculate the variance of the jth subproblem var_j.
7:          Calculate the difference between the two subproblems $dis_j = |\bar{g}^i - \bar{g}^j|$, where $\bar{g}^i$ and $\bar{g}^j$ are estimated objective values of the ith and jth subproblems.
8:          $B_2(i) = B_2(i) \backslash \{j\}$.
9:      **end while**
10:     Use the OCBA technique to compute the sample size $k_j (j = 1, \ldots, |B_2(i)|)$ for each neighboring solution of the ith subproblem.
11:     **for** $j = 1$ to $|B_2(i)|$ **do**
12:         $q_j = q_j + k_j$.
13:     **end for**
14: **end for**

Actually, the strength levels of noise cannot be known in advance and are usually estimated by variance values of individuals [Rakshit and Konar (2015b)]. Specifically, the estimated noise level in the ith objective function is defined as follows:

$$\hat{\sigma}_i^2 = \sum_{j=1}^{N} \sigma_{ij}^2 \bigg/ (N \cdot |\widehat{F_i^{\max}} - \widehat{F_i^{\min}}|), \tag{5.1}$$

where σ_{ij}^2 represents the variance value in the ith objective function of the jth solution, and N is the number of solutions in the population. $\widehat{F_i^{\max}}$ and $\widehat{F_i^{\min}}$ denote the maximum and minimum value of the ith objective function in the population, respectively. Besides, no prior information is required while estimating the noise level.

5.2.3.1 *Mixed objective evaluation strategy*

As can be seen from Fig. 5.4, DMOEAs with the mean criterion show better convergence performance, and DMOEAs adopting the median criterion maintain better diversity of the evolving population. Results on different test problems display high similarity. We define a parameter, denoted as r_{median}, to represent the percentage of the median value in the fitness estimation of a solution. In other words, the fitness estimation of a solution is the sum of $r_{median} \cdot 100\%$ of the median value of samples and $(1 - r_{median}) \cdot 100\%$ of the mean value of samples. The parameter r_{median} varies according to the following equation:

$$r_{median} = \begin{cases} 0 & \hat{\sigma}^2 < 1\%, \\[2ex] r_1 + \Delta_1 \cdot \dfrac{\hat{\sigma}^2}{\max(\hat{\sigma}^2)} & 1\% \leq \hat{\sigma}^2 < 10\%, \\[2ex] r_2 + \Delta_2 \cdot \dfrac{\hat{\sigma}^2}{\max(\hat{\sigma}^2)} & \hat{\sigma}^2 \geq 10\%, \end{cases} \tag{5.2}$$

where $\hat{\sigma}^2$ is the averaged estimated noise level of all objective functions, i.e., $\hat{\sigma}^2 = \sum_{i=1}^{m} \hat{\sigma}_i^2 / m$, where m is the number of objectives. $\max(\hat{\sigma}^2)$ returns the maximum of estimated noise levels found so far. r_i and Δ_i $(i = 1, 2)$ are random values in the interval $[0, 1]$.

When the estimated noise level is low (i.e., lower than 1%), DMOEAs with mean and median criteria perform similarly. Since the mean value is easier to compute, the mean value is regarded as the final fitness estimation. When the estimated noise level increases, we integrate $r_i\% \sim (r_i + \Delta_i)\%$ $(i = 1, 2)$ of the median value into the objective estimation. Based on empirical studies, these parameters are set as $r_1 = 0.3$, $r_2 = 0.1$ and $\Delta_i = 0.2$ $(i = 1, 2)$. Specifically, when the estimated noise level is medium (i.e., falls between 1% and 10%), 30% $\sim$ 50% of the median value of a series of samples is integrated into the objective estimation. This attempts to enhance the diversity of population and ensure convergence at the same time. When the estimated noise level is high (i.e., higher than or equal to 10%), the final fitness estimation consists of 70% $\sim$ 90% of the mean value, since it is easy to maintain the diversity in the presence of high-level noise.

5.2.3.2 *Mixed repair mechanism*

As mentioned before, it is obvious that the commonly used truncation repair mechanism gives rise to the loss of diversity in the decision space, which also results in losing diversity in the objective space. The proposed mixed repair mechanism adopts the truncation repair mechanism and the random repair mechanism simultaneously. Similar to the mixed objective evaluation strategy, a parameter that represents the percentage of infeasible variables that use the random repair mechanism is defined and denoted as r_{random}. Thus, when repairing an infeasible solution, $\lceil r_{random} \cdot 100\% \rceil$ of infeasible variables will be repaired by the random repair mechanism, where $\lceil \cdot \rceil$ returns the nearest integer in the direction of positive infinity. The remaining infeasible variables will be repaired via the truncation repair mechanism. The parameter r_{random} changes according to the following equation:

$$
r_{random} = \begin{cases} 0 & \hat{\sigma}^2 < 1\% \\[2ex] r_3 + \Delta_3 \cdot \dfrac{\hat{\sigma}^2}{\max(\hat{\sigma}^2)} & 1\% \leq \hat{\sigma}^2 < 10\% \\[2ex] r_4 + \Delta_4 \cdot \dfrac{\hat{\sigma}^2}{\max(\hat{\sigma}^2)} & \hat{\sigma}^2 \geq 10\%. \end{cases} \tag{5.3}
$$

Similar to the previous subsection, these parameters are given as $r_3 = 0.5$, $r_4 = 0.3$ and $\Delta_i = 0.2$ ($i = 3, 4$), according to empirical results. Specifically, the truncation repair mechanism is still adopted when the estimated noise is small since low-level noise has little effect on the performance of DMOEAs on UMOPs. When the estimated noise level is medium, we apply the random repair mechanism on $50\% \sim 70\%$ of infeasible variables. Finally, when the estimated noise level is high, only $30\% \sim 50\%$ of the infeasible variables will be repaired via the random repair mechanism.

Given the above, the proposed mixed objective evaluation strategy and the mixed repair mechanism are trade-offs between two evaluation criteria and two repair mechanisms, respectively. They adaptively strike a balance between convergence and diversity from the aspects of objective space and decision space, respectively.

5.2.4   *Implementation*

These proposed techniques are incorporated into DMOEAs, and the resulting noise-tolerant DMOEAs are named NT-DMOEAs. Two representative DMOEAs, i.e., MOEA/D and DMOEA-εC, are taken as examples. MOEA/D and DMOEA-εC with these noise-tolerant features are named as NT-MOEA/D and NT-DMOEA-εC, respectively. The notations used in the NT-DMOEAs are given in Table 5.1. The algorithmic description of NT-DMOEAs is presented in **Algorithm 15**.

Algorithm 15 Framework of NT-DMOEAs

Require: A UMOP and related parameters.

Ensure: An external archive population EP.

1: **for** i=1 to N **do**
2:    Set two types of neighborhoods for the ith subproblem, i.e., $B(i)$ and $B_2(i)$.
3: **end for**
4: Initialize samples of noise randomly.
5: Randomly initialize the evolving population $P = \{x^1, \ldots, x^N\}$; apply the mixed objective evaluation strategy to calculate sampled objective values of each subproblem, initialize the ideal point z^*.
6: Set $gen = 0$ and $nfe = N$.
7: **while** $nfe \leq NFE\_preprocess$ **do**
8:    Apply NSGA-II to obtain the population P'.
9:    $nfe = nfe + N \cdot n$; $gen = gen + 1$.
10: **end while**
11: Extract non-dominated individuals from P' and denote the set of them as EP; estimate the nadir point z^{nad} and take P' as the initial population for the following DMOEAs.
12: **while** $nfe \leq (NFE - NFE\_preprocess)$ **do**
13:    **if** gen is a multiple of $DRA\_interval$ **then**
14:       Update the indices of the subproblems I that will be processed in the next generation by applying the dynamic resource allocation scheme [Zhang et al. (2009a)].
15:    **end if**
16:    **for** $i \in I$ **do**
17:       $P = \begin{cases} B_1(i), & if\ rand < \delta \\ \{1, 2, \ldots, N\}, & otherwise \end{cases}$
18:       Reproduction: Select parent individuals from P' randomly and then apply a certain reproduction operator to generate a new solution y and reevaluate the new solution k_1 times.
19:       Repair: If y is infeasible, use the mixed repair mechanism to repair it.
20:       Using adaptive sampling strategies (**Algorithm 13** or **Algorithm 14**) to allocate the remaining $(n - k_1)$ samples among neighboring solutions in $B_2(i)$.
21:       Apply the mixed objective evaluation strategy to update objective estimation of each neighboring solution; $nfe = nfe + n$.
22:       Update the approximated ideal point z^*.
23:       Compare y with neighboring solutions in $B_1(i)$ and update these neighboring solutions by using the objective values directly or the feasibility rule.
24:       Update the external archive EP and prune it by using the crowding distance approach.
25:       Update the approximated nadir point z^{nad}.
26:    **end for**
27:    $gen = gen + 1$.
28: **end while**

Table 5.1   Summary of Notations Used in the Description of NT-DMOEAs.

N	The number of weight vectors or upper bound vectors (the same as the population size)
n	Predefined sample size for each subproblem
k_1	Predefined sample size for a newly generated solution
T_1	Neighborhood size for the update of solutions
T_2	Neighborhood size for the update of samples
δ	Probability of selecting mate solutions from its neighborhood
$NFE\_preprocess$	The number of function evaluations used by Pareto-based MOEAs
$DRA\_interval$	Iteration interval of utilizing the dynamic resource allocation strategy
NFE	Maximum number of function evaluations
$rand$	A randomly distributed value in the interval $[0,1]$

5.3   Numerical Experiments

This section first conducts preliminary experiments of algorithmic performances on parameter settings and then performs two types of numerical experiments. The first one is to compare relative performances of the individual extensions embedded in NT-DMOEAs, and the second one is to compare NT-DMOEAs with other noise-tolerant algorithms. Seven different noise strength levels of $\sigma^2 \in \{0.1\%, 0.2\%, 0.5\%, 1\%, 5\%, 10\%, 20\%\}$ are still applied to all test instances. Due to the differences in algorithmic frameworks, N is set to 351, 324 and 300 for the DTLZ instances for MOEA/D, DMOEA-εC and the remaining algorithms, respectively. The neighborhood sizes for the update of solutions and the update of samples are set as $T_1 = \lfloor 10\% \cdot N \rfloor$ and $T_2 = \lfloor 10\% \cdot N \rfloor$, respectively. The sample size of each subproblem is set as $n = 5$, and the number of samples for each new solution is set as $k_1 = 2$, based on the parameters tuning, which will be presented in the following. Moreover, other parameters are the same as those mentioned before. All comparison algorithms terminate when the number of evaluations reaches the maximum number.

5.3.1   *Parameters Tuning*

In this section, the parameters analyses of NT-DMOEAs are deeply researched. Specifically, effects of the parameters T_1, T_2, n and k_1 on the performances of NT-DMOEAs are investigated.

5.3.1.1   Effects of T_1 and T_2

In NT-DMOEAs, the neighbourhood sizes T_1 and T_2 are used for the purposes of updating neighboring solutions and updating samples, respectively. We investigate how the NT-DMOEA-εC is influenced by the combination of two parameters. We take ZDT2 and UF2 as examples and test different settings of T_1 and T_2 in the implementation of NT-DMOEA-εC. Different T_1 and T_2 values are both set as $\lfloor (3\%, 5\%, 10\%, 30\%, 50\%) \cdot N \rfloor$. All the other parameters are kept the same as described before, and 30 independent runs have been conducted for each configuration on these test instances.

Figure 5.5 shows the mean Δ_p metric values across all combinations of T_1 and T_2 values on the selected test problems. As shown in Fig. 5.5,

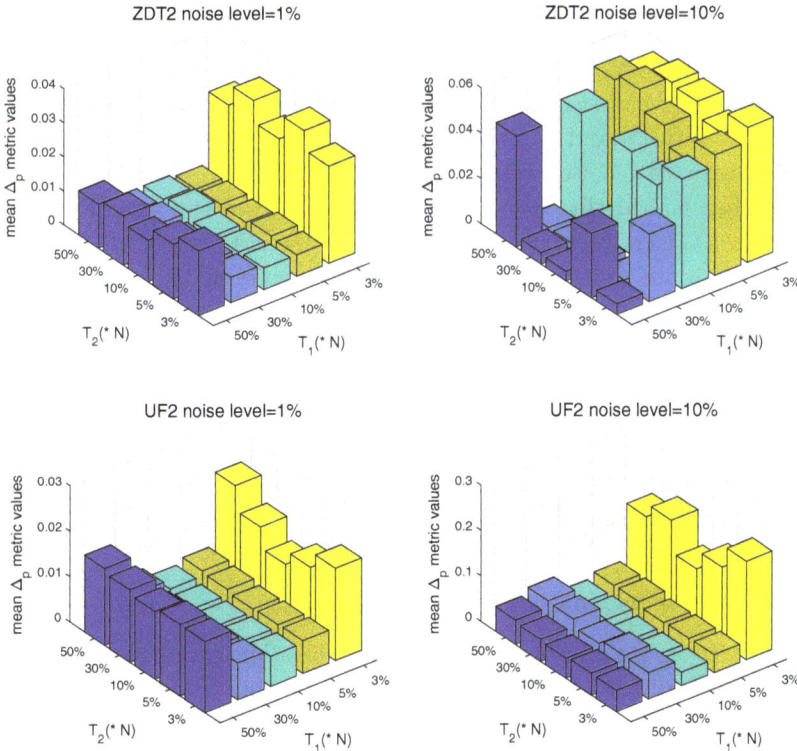

Fig. 5.5   Mean Δ_p values within 30 runs versus values of T_1 and T_2 in NT-DMOEA-εC for ZDT2 and UF2 test problems with 1% and 10% noise levels, respectively.

NT-DMOEA-εC performs well with a wide range of T_1 and T_2 values on both ZDT2 and UF2 instances, except ZDT2 with a 10% noise level on which a large T_1 performs better. For the majority of test problems, $T_1 = \lfloor 10\% \cdot N \rfloor$ and $T_2 = \lfloor 10\% \cdot N \rfloor$ are the best parameter settings.

5.3.1.2   *Effects of n and k_1*

Here, n and k_1 are major parameters in the sampling strategy, and they decide the sample size for each solution and the number of samples for each newly generated solution, respectively. Similarly, we take ZDT2 and UF2 as examples and study the effects of the different settings of n and k_1 in NT-MOEA/D. Different n and k_1 values are set as $\{3, 4, 5, 7, 9\}$ and $\{2, 3, 4, 5, 6\}$, respectively. Figure 5.6 displays the mean Δ_p metric values across all combinations of n and k_1 values on the selected test problems. It should be noted that for the combination that $k_1 > n$, we set $k_1 = n$.

As can be seen in Fig. 5.6, NT-MOEA/D performs well with $n \in \{3, 4, 5\}$ for ZDT2 with 1% and 10% noise levels and UF2 with a 10% noise level. As to UF2 with a 1% noise level, NT-MOEA/D performs well when $n \in \{4, 5\}$. It can be claimed that $n = 5$ is a good setting for the majority of test instances. Generally, a larger n is good for the stability of the proposed algorithm, while a smaller n gives a high probability of detecting new solutions.

Figure 5.6 reveals that NT-MOEA/D with $k_1 \in \{2, 3, 4, 5, 6\}$ values shows very different performances on both problems, especially on UF2 with a 10% noise level. In order to allocate more samples among neighborhoods based on the information of the current population, we set $k_1 = 2$, with which NT-MOEA/D exhibits competitive performances on both test instances with various levels of noise.

5.3.2   *Comparisons with Various Variants*

In this section, the effects of proposed noise-handling techniques are deeply analyzed on all test problems with different strength levels of noise. First, effects of the Pareto-based nadir point estimation strategy (PNE) and two adaptive sampling strategies (ASs) are examined. Then, the effects of the mixed objective evaluation strategy (MO) and the mixed repair mechanism (MR) are further investigated.

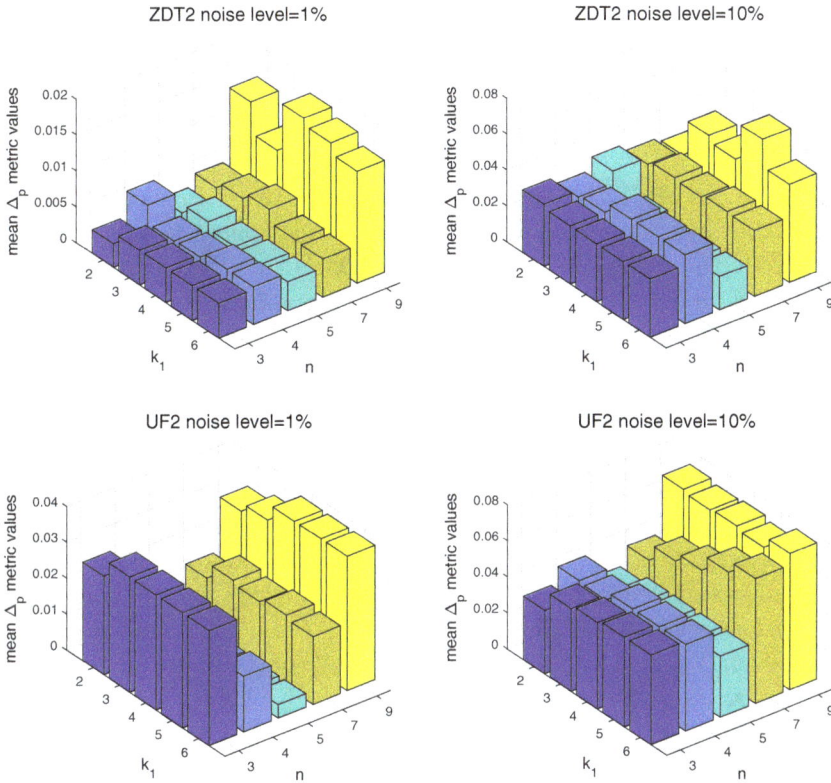

Fig. 5.6   Mean Δ_p values within 30 runs versus values of n and k_1 in NT-MOEA/D for ZDT2 and UF2 test problems with 1% and 10% noise levels, respectively.

5.3.2.1   *Effects of the Pareto-based nadir point estimation strategy and adaptive sampling strategies*

As mentioned above, the Pareto-based nadir point estimation strategy takes full advantage of the high speed of convergence at the early stage of optimization of the Pareto-based MOEAs. However, does this nadir point estimation strategy indeed enhance the performances of standard DMOEAs (SDMOEAs)?[2] In order to answer this question, a variant of SDMOEAs, denoted as SDMOEAs-PNE, is developed for comparing with SDMOEAs.

---

[2]SDMOEAs mean standard DMOEAs adopting the uniform sampling strategy, the mean value of samples as the objective estimation, and the truncation repair mechanism.

Furthermore, do two types of adaptive sampling strategies, i.e., co-ranking and OCBA strategies, play an important role in SDMOEAs? Similarly, two SDMOEAs variants, denoted as SDMOEAs-PNE-CR and SDMOEAs-PNE-OCBA are designed to answer this question. Detailed descriptions of these variants are given in the following.

SDMOEAs-PNE: Different from SDMOEAs, the process of estimating the nadir point is replaced with the Pareto-based approaches.

SDMOEAs-PNE-CR: Different from SDMOEAs-PNE, the number of samples is adaptively allocated according to the co-ranking allocation strategy (**Algorithm 13**).

SDMOEAs-PNE-OCBA: Different from SDMOEAs-PNE, the number of samples is allocated via the OCBA strategy (**Algorithm 14**).

All three variants are experimentally compared with SDMOEAs on all test instances with seven different strength levels of noise. The mean rank values in terms of the Δ_p metric over 30 independent runs of SDMOEA-εC and its variants on 4 test instances with 7 different strength levels of noise are shown in Table 5.2. The bold data in the table are the best mean rank values for each instance. Besides, the last row of each table summarizes the overall performance, i.e., the mean ranks of all compared algorithms on all test instances with seven different strength levels of noise in terms of the Δ_p metric values.

As can be seen in Table 5.2, in terms of Δ_p metric values, SDMOEA-εC-PNE shows significant advantages over SDMOEA-εC on all test problems. The performances of SDMOEA-εC-PNE with two types of adaptive sampling strategies, i.e., SDMOEA-εC-PNE-CR and SDMOEA-εC-PNE-OCBA are better than that of SDMOEA-εC and SDMOEA-εC-PNE. Besides, SDMOEA-εC-PNE-CR shows superiority

Table 5.2   Overall Performances of SDMOEA-εC and its Variants about PNE and 2 ASs over 30 Independent Runs on 4 Test Instances with 7 Different Strength Levels of Noise in Terms of Δ_p Metrics.

Instance	SDMOEA-εC	-PNE	-PNE-CR	-PNE-OCBA
ZDT2	3.856	2.571	**1.429**	2.143
DTLZ2	4.000	2.857	**1.000**	2.143
UF2	3.571	3.143	**1.428**	1.857
UF10	3.571	3.143	**1.429**	1.857
Mean Rank	3.714	2.950	**1.420**	1.706

over SDMOEA-εC-PNE-OCBA on all test problems in terms of Δ_p metrics. As to the MOEA/D, similar results can be obtained with respect to Δ_p metrics. Specifically, SMOEA/D-PNE with two types of adaptive sampling strategies outperforms SMOEA/D-PNE whose performance is better than SMOEA/D. Among two adaptive sampling strategies, the co-ranking allocation strategy shows better performance than the OCBA strategy. Results in this subsection highlight the effectiveness of the Pareto-based nadir point estimation strategy and two adaptive sampling strategies, especially the co-ranking allocation strategy, embedded in both DMOEA-εC and MOEA/D with respect to Δ_p metric values. Since SDMOEAs with the Pareto-based nadir point estimation strategy and the co-ranking adaptive sampling strategy perform best among their variants, we take them as baselines for further investigation and denote them as SDMOEAs for short.

5.3.2.2 *Effects of the mixed objective estimation strategy and the mixed repair mechanism*

Experiments in this subsection are conducted in a similar manner as in the previous subsection with an aim to compare SDMOEA and its variants and show superiority of the proposed mixed Objective evaluation strategy and the mixed repair mechanism. What's more, since the proposed two mechanisms are applicable to NSGA-II, similar experiments are performed to verify the effectiveness of two strategies in NSGA-II. In order to further investigate the superiority of two mixed objective strategies, we develop the following variants.

-ME: Different from SDMOEAs and NSGA-II, the median value is regarded as the objective estimation.

-MO: In this variant, the mixed objective evaluation is adopted as the objective value.

-RR: Different from SDMOEAs and NSGA-II, the truncation repair mechanism is replaced with the random repair method after an infeasible solution is generated.

-MR: In this variant, the mixed repair mechanism is adopted after an infeasible solution is generated.

-MM: In this variant, the mixed objective evaluation is regarded as the objective value and the mixed repair mechanism is adopted after an infeasible solution is generated.

Table 5.3   Overall Performances of SMOEA/D and its Variants about MO and MR over 30 Independent Runs on 4 Test Instances with 7 Different Strength Levels of Noise in Terms of Δ_p Metrics.

Instance	SMOEA/D	-ME	-MO	-RR	-MR	-MM
ZDT2	5.143	3.714	2.571	4.143	3.286	**2.143**
DTLZ2	5.571	4.714	3.143	3.143	2.429	**2.000**
UF2	5.286	3.857	2.000	5.000	3.143	**1.857**
UF10	5.429	3.429	2.429	4.143	3.714	**1.857**
Mean Rank	5.170	3.781	2.840	3.781	3.025	**2.076**

Table 5.4   Overall Performances of NSGA-II and its Variants About MO and MR over 30 Independent Runs on 4 Test Instances with 7 Different Strength Levels of Noise in Terms of Δ_p Metrics.

Instance	Reference Point					
	NSGA-II	-ME	-MO	-RR	-MR	-MM
ZDT2	5.561	3.143	2.429	5.143	3.143	**1.571**
DTLZ2	5.426	3.286	2.714	4.714	3.286	**1.571**
UF2	5.286	2.857	2.429	5.143	3.286	**2.000**
UF10	5.857	2.857	2.571	4.714	3.143	**1.714**
Mean Rank	5.478	3.076	2.555	5.059	3.017	**1.807**

All five variants are experimentally compared with SDMOEAs and NSGA-II on all test instances with seven different strength levels of noise. Tables 5.3 and 5.4 display the mean rank values in terms of the Δ_p metric over 30 independent runs of SMOEA/D and NSGA-II and its variants on 4 test instances, respectively. The bold data in each table are the best mean rank values for each instance. The last row of each table summarizes the overall performance, i.e., the mean ranks of all compared algorithms on all test instances with seven different strength levels of noise in terms of the Δ_p metric.

As can be seen in Tables 5.3 and 5.4, in terms of Δ_p metric values, SMOEA/D-MO shows a significant advantage over SMOEA/D and SMOEA/D-ME on all test problems. SMOEA/D-MR outperforms the SMOEA/D and SMOEA/D-RR on all test instances in terms of Δ_p metric values. Besides, it is observed that SMOEA/D-MM performs best among all variants on all test problems in terms of Δ_p metrics. As to NSGA-II, similar results can be obtained with respect to the Δ_p metric.

Specifically, NSGA-II-MO and NSGA-II-MR both show superiority over NSGA-II, and they outperform NSGA-II-ME and NSGA-II-RR, respectively. Also, NSGA-II-MM shows the best performance compared with other variants on all test problems in terms of Δ_p metrics. Tables 5.3 and 5.4 demonstrate that DMOEAs-MM, denoted as NT-DMOEAs in the following, performs better than all the other variants on all test instances in terms of the Δ_p metric values. Besides, NSGA-II-MM shows advantages over its variants on all test instances, and it will be adopted as a competitor in the following comparison experiments. Therefore, the effectiveness of the mixed objective evaluation strategy and the mixed repair mechanism are confirmed experimentally.

5.3.3 *Comparisons with Other Noise-Tolerant Algorithms*

In order to examine the effectiveness of NT-DMOEAs, including the NT-DMOEA-εC and NT-MOEA/D, a comparative study with the NSGA-II-MM, BES [Fieldsend and Everson (2005)], MOP-EA [Syberfeldt *et al.* (2010)],[3] and RTEA [Fieldsend and Everson (2015)] is carried out based upon all benchmark problems with seven different strength levels of noise.

Table 5.5 exhibits the mean rank values in terms of both Δ_p and HV metrics over 30 independent runs for each test instance with 7 different strength levels of noise in order to have a global view of performance of all algorithms. The bold data in the table are the best mean rank values for each instance. Besides, the last row of Table 5.5 summarizes the overall performance, i.e., the mean ranks of five algorithms on all instances in terms of both metric values.

As can be seen in Table 5.5, in terms of HV metric values, NT-DMOEA-εC and NT-MOEA/D show significant advantages over comparison algorithms on the majority of 17 test instances except DTLZ4, UF9 and UF10, on which RTEA shows better or similar performance. As to the remaining algorithms, RTEA outperforms other algorithms remarkably. What's more, NSGA-II-MM and MOP-EA perform similarly on the majority of test problems and exhibit obvious advantages over BES. To be specific, NT-DMOEA-εC has the lowest rank value on ZDT2, ZDT3 and ZDT6 test

---

[3]The algorithm of Syberfeldt *et al.* (2010) is designed to cope with variable noise. We follow the practice of Syberfeldt *et al.* (2010) and do not incorporate the surrogate, which is designed for expensive real-world problems. We label the algorithm MOP-EA here.

Table 5.5 Overall Performances of 6 Algorithms over 30 Independent Runs on 17 Instances with 7 Different Strength Levels of Noise in Terms of Δ_p and HV Metrics.

Instance	NT-DMOEA-εC Δ_p	HV	NT-MOEA/D Δ_p	HV	NSGA-II-MM Δ_p	HV	BES Δ_p	HV	MOP-EA Δ_p	HV	RTEA Δ_p	HV
ZDT1	**1.857**	**2.000**	2.000	2.143	4.714	4.857	5.571	5.429	4.571	4.571	2.286	**2.000**
ZDT2	**2.000**	**1.857**	2.143	2.429	4.429	4.429	5.714	5.571	4.429	4.429	2.286	2.286
ZDT3	**1.714**	**1.571**	1.857	2.143	4.857	4.857	5.429	5.571	4.714	4.571	2.429	2.286
ZDT4	**2.000**	**1.857**	**2.000**	**1.857**	4.571	4.429	5.286	5.429	5.000	5.000	2.143	2.429
ZDT6	1.714	**1.714**	2.143	2.429	4.714	4.429	5.571	5.714	4.714	4.857	2.143	1.857
DTLZ2	2.000	**2.000**	**1.857**	**2.000**	4.857	5.000	4.857	5.000	5.143	4.143	2.286	2.857
DTLZ4	**1.714**	**1.857**	2.286	2.429	4.429	4.571	5.571	5.714	5.000	4.714	2.000	1.714
UF1	2.143	1.857	**1.714**	**1.714**	4.714	4.571	5.286	5.429	4.857	4.714	2.286	2.714
UF2	**1.714**	**1.714**	2.000	1.857	4.714	4.714	5.286	5.286	4.714	4.714	2.571	2.714
UF3	**1.857**	**1.714**	2.286	2.429	4.571	4.429	5.429	5.714	4.857	4.429	2.000	2.286
UF4	**1.571**	**1.714**	2.143	1.857	4.714	4.857	5.571	5.429	4.714	4.714	2.286	2.429
UF5	2.143	2.000	**2.000**	**1.857**	4.857	4.857	4.857	4.857	4.857	5.143	2.286	2.286
UF6	**1.857**	**1.857**	**1.857**	2.143	4.714	4.571	5.571	5.571	4.571	4.857	2.429	2.000
UF7	2.286	2.143	**2.143**	**2.000**	4.857	4.714	4.857	5.000	4.714	4.571	2.143	2.571
UF8	**1.714**	1.857	1.857	**1.714**	4.714	4.857	5.714	5.857	4.571	4.286	2.429	2.429
UF9	**2.000**	2.286	2.429	2.429	4.714	4.571	5.429	5.571	4.286	4.000	2.143	**2.143**
UF10	**2.000**	**2.286**	2.286	2.286	4.714	4.571	5.000	5.143	4.714	4.143	2.286	**2.286**
Mean Rank	**1.899**		2.050		4.681		5.391		4.647		4.572	

problems. RTEA and NT-DMOEA-εC have the same rank values and show superiority over other algorithms on ZDT1. As to ZDT2 and ZDT4, NT-DMOEA-εC and NT-MOEA/D both have the lowest rank value, which means the best performance among all comparison algorithms. For UF1, UF5, UF7 and UF8 test problems, NT-MOEA/D performs better than NT-DMOEA-εC and RTEA. As to UF2, UF3, UF4 and UF6 instances, NT-DMOEA-εC shows better rank values over NT-MOEA/D and RTEA. Similar results can be obtained in terms of Δ_p metric values according to statistical results in Table 5.5. That is, NT-DMOEA-εC, NT-MOEA/D and RTEA outperform other comparison algorithms on all 17 test instances. NT-DMOEA-εC and NT-MOEA/D have smaller rank values than RTEA on the majority of test instances. To summarize, as can be seen in Table 5.5, NT-DMOEA-εC and NT-MOEA/D perform better than other comparison algorithms on all test problems. NT-DMOEA-εC achieves the best *HV* and Δ_p metric values over others on the majority of test instances except DTLZ4, UF9 and UF10, on which RTEA exhibits better or similar performances.

The last row of Table 5.5 summarizes these statistical results and reveals the overall rank of the six algorithms, that is, NT-DMOEA-εC, NT-MOEA/D, RTEA, MOP-EA, NSGA-II-MM and BES, according to the mean rank values of both Δ_p and *HV* metric values. It indicates that NT-DMOEA-εC and NT-MOEA/D have advantages over comparison algorithms on all test instances and NT-DMOEA-εC shows the best performance on these test problems in terms of two metrics. The superiority of NT-DMOEA-εC and NT-MOEA/D can be attributed to the Pareto-based nadir point estimation strategy, two adaptive sampling strategies, the mixed objective estimation strategy, and the mixed repair mechanism.

Figure 5.7 shows the distribution of the final solutions with the minimum Δ_p value of 4 test instances with 0.1%, 1% and 10% noise levels within 30 runs found by NT-DMOEA-εC. It is visually evident that for ZDT and the bi-objective UF with a 0.1% noise level, the final population obtained by NT-DMOEA-εC can cover the whole PFs very well and spread uniformly. NT-DMOEA-εC also shows good convergence and obtains solutions with good diversity on ZDT and bi-objective UF with a 1% noise level. For ZDT and the bi-objective UF with a 10% noise level, final solutions obtained by NT-DMOEA-εC do not approximate the PFs very well but spread widely along the PFs. As to the DTLZ and tri-objective UF instances with noise, even the 0.1% noise level has a significant impact on final solutions obtained

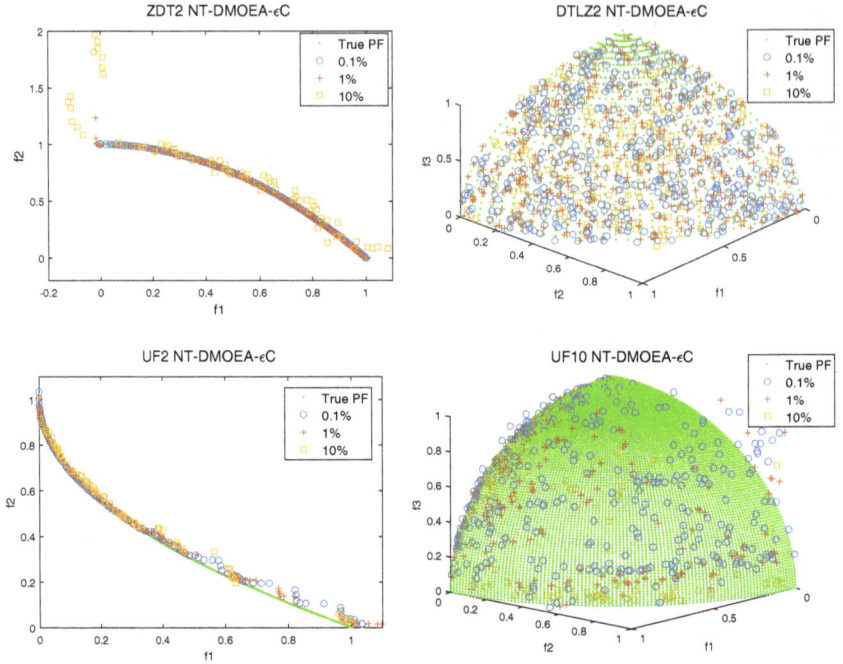

Fig. 5.7   Final populations in the objective space with the minimum Δ_p metric value within 30 runs obtained by NT-DMOEA-εC on 4 test problems with 0.1%, 1%, and 10% noise levels.

via NT-DMOEA-εC. Specifically, NT-DMOEA-εC can only find solutions that are not approximated nearly enough to the true PF but spread widely along the true PF. For DTLZ and the tri-objective UF with 1% and 10% noise levels, both convergence and diversity of the final solutions obtained via NT-DMOEA-εC deteriorates. In summary, Fig. 5.7 shows that NT-DMOEA-εC can achieve the approximations with both good convergence and diversity for most of ZDT and bi-objective UF test instances, and can find solutions with good diversity for DTLZ and tri-objective UF test instances.

5.4   Conclusion

Uncertainties widely exist in real-world problems, which are mostly with multiple conflicting objectives. Usually the uncertainty is modeled as additive noise in the objective space. The decomposition-based multi-objective evolutionary algorithms (DMOEAs) are important approaches for multi-objective optimization and have not been applied to uncertain problems extensively. In order to design effective noise-tolerant DMOEAs (NT-DMOEAs) for uncertain problems, this chapter has conducted extensive studies to examine the impact of noise on DMOEAs, particularly for the population dynamics of convergence and diversity. Based on observations of noise impacts on DMOEAs, four noise-handling techniques have been proposed and embedded in two popular DMOEAs, i.e., MOEA/D and DMOEA-εC. The proposed NT-DMOEAs utilize the composite benefits of four major extensions.

Specifically, the Pareto-based nadir point estimation strategy makes full use of the high speed of convergence at the early stage of optimization of the Pareto-based MOEAs to improve the accuracy of nadir point estimation. Two adaptive sampling strategies vary the number of samples used per solution based on the differences among neighboring solutions and their variance to control the trade-off between search space exploration and exploitation. Then the mixed objective evaluation strategy varies the estimation of an uncertain objective function between the mean and median values of fitness samples adaptively, according to the estimated strength level of noise to alleviate effects of noise. Finally, the mixed repair mechanism adaptively combines the truncation repair mechanism and a random repair mechanism together to adaptively remedy the loss of diversity in the decision space.

Two NT-DMOEAs have been compared with their various variants and four noise-tolerant algorithms to show the superiority of proposed features on 17 test instances with different strength levels of noise. A systematical experimental study has shown that two NT-DMOEAs significantly outperform their variants on the majority of test instances with different strength levels of noise. It also has shown that two NT-DMOEAs, especially NT-DMOEA-εC, exhibit competitive or superior performance in terms of proximity and diversity for the majority of test instances.

Many real-world problems suffer from noisy disturbances. Under such circumstances, standard DMOEAs are inadequate to provide Pareto optimal solutions for these noisy problems, which calls for the necessity of noise-tolerant algorithms. Actually, it is usually difficult to give a good representation of noise in real environments, thus estimating the noise accurately is an important task when handling noisy optimization problems. The most common way of estimating noise is to sample candidate solutions multiple times and measure the strength level of noise based on the variance value of a series of samples. This method is simple and straightforward, but not effective enough when the characteristics of noise are difficult to describe. What's more, there are many different forms of noise whose characteristics change as a function of location (in the design or objective space), or which alter during the course of an optimization [Fieldsend and Everson (2015)]. This poses new challenges to the uncertain optimization. It will be our future work to propose an effective way of giving a good representation of the noise and estimating the strength level of noise accurately. Furthermore, new mechanisms that can deal with other types of uncertainties will be investigated. The performances of proposed NT-DMOEAs will also be tested on more complex real-world problems.

Chapter 6

The Bi-objective Critical Node Detection Problem with Minimum Pairwise Connectivity and Cost: Theory and Algorithms

An effective way to analyze and apprehend structural properties of networks is to find their most critical nodes. This chapter studies a bi-objective critical node detection problem, denoted as Bi-CNDP. In this variant, we do not make any assumptions on the psychology of decision makers and seek to find a set of solutions that minimizes the pairwise connectivity of the induced graph and the cost of removing these critical nodes at the same time. After explicitly stating the formulation of the Bi-CNDP, we first prove the NP-hardness of this problem for general graphs and the existence of a poly-nomial algorithm for constructing the ε-approximated Pareto front (PF) for Bi-CNDPs on trees. Then, different approaches of determining the mating pool and the replacement pool are proposed for the decomposition-based multi-objective evolutionary algorithms (MOEAs). Based on this, two types of decomposition-based MOEAs (i.e., MOEA/D and DMOEA-εC) are mod-ified and applied to solve the proposed Bi-CNDP. Numerical experiments on 16 famous benchmark problems with random and logarithmic weights are firstly conducted to assess different types of the mating pool and the replacement pool. Besides, computational results between two improved algorithms, i.e., I-MOEA/D and I-DMOEA-εC, demonstrate that they behave differently on these instances and I-DMOEA-εC shows better perfor-mance on the majority of test instances. Finally, a decision-making process from the perspective of minimizing the pairwise connectivity of the induced graph given a constraint on the cost of removing nodes is presented for help-ing decision makers identify the most critical nodes for further protection or attack.

6.1   Introduction

Common networks such as telecommunication, transportation, power systems and others are exposed to various threats coming from the environment. Any failure of elements of these networks may lead to a complete or partial halt of their services and result in unexpected consequences [Atputharajah and Saha (2009); Liang *et al.* (2017)]. An effective way to analyze and apprehend structural properties of networks is to find the most critical nodes. A node is critical if its failure or removal significantly degrades the performance of a network. Once identified, the critical node can be used to either implement protection or attack strategies for networks. The identification of critical nodes in various networks is a fundamental task. In the literature, this problem has attracted a significant amount of research attention in a number of fields, including social network analysis [Borgatti (2006); Fan and Pardalos (2010); Kempe *et al.* (2010); Leskovec *et al.* (2007)], transportation networks' vulnerability assessment [Jenelius *et al.* (2006)], power grid construction [Salmeron *et al.* (2015)], network risk management [Arulselvan *et al.* (2007)], epidemic control [Zhou *et al.* (2006)], and network immunization strategies [Arulselvan *et al.* (2009); Kuhlman *et al.* (2010)]. Besides, a number of applications in the military domain have also been researched [Walteros and Pardalos (2012)].

In this chapter, we are motivated primarily by the critical node detection problem (CNDP) described in [Arulselvan *et al.* (2009)]. Given a network, the CNDP consists in finding a set of nodes, the deletion of which results in minimizing a connectivity measure in the induced graph. Different connectivity measures can be devised according to specific applications of interest. The choice of different measures typically leads to different optimal solutions, as described in [Aringhieri *et al.* (2016a); Shen and Smith (2012); Veremyev *et al.* (2014b)]. In the literature, three mostly common used connectivity measures are: 1) pairwise connectivity, i.e., the number of pair of nodes connected by a path inside the graph; 2) the size of the largest connected component; and 3) the number of connected components [Aringhieri *et al.* (2016a)]. Even though these measures are different and can lead to different optimal solutions, they are not totally unrelated. The ideal situation for minimizing the pairwise connectivity is to obtain the largest number of connected components with the smallest variance in their cardinality. This implies that minimizing the pairwise connectivity is a trade-off between minimizing the cardinality of the largest connected component and maximizing the number of connected components. This chapter concentrates on the pairwise connectivity measure, which is generally enough to determine which nodes are still connected in the induced network.

Considering one of the above-mentioned connectivity measures, several variants of the CNDP have been investigated. Among them, the first two variants, namely CNP and CC-CNP, are the most commonly used formulation in the literature. Firstly, the basic CNDP (denoted as CNP in the original [Arulselvan *et al.* (2009); Tomaino *et al.* (2012)]) aims at minimizing the connectivity measure in the induced graph, given a constraint on the maximum number of nodes that can be removed. Secondly, in the cardinality constrained critical node detection problem (CC-CNP) [Arulselvan *et al.* (2011)], the objective is to minimize the number of nodes required to be removed given the maximum size of connected components. It belongs to the connectivity constrained formulation that minimizes the number of deleted nodes in order to meet a threshold on a certain connectivity metric. The CNDPs are also related to a variety of graph fragmentation problems in the literature. The vertex separator problem [Balas and Souza (2005)] and the *k*-separator problem [Ben-Ameur *et al.* (2015)] have the most in common with the CNDP. Other relevant examples include the minimum contamination problem [Kumar *et al.* (2010)], the sum-of-squares partitioning problem [Aspnes *et al.* (2005); Chen *et al.* (2010)], and so on. Further information pertaining to graph fragmentation problems can be found in [Shen and Smith (2012)].

Theoretical studies on combinatorial aspects of the CNDP for general graphs were first given in [Arulselvan *et al.* (2009); Di Summa *et al.* (2011)]. As mentioned above, Arulselvan and colleagues presented two variants of the CNDP, namely CNP [Arulselvan *et al.* (2009)] and CC-CNP [Arulselvan *et al.* (2011)]. They also proved the complexity of two variants for general graphs and introduced some heuristic algorithms for solving them. Since then, many studies have been presented depending on different connectivity metrics to be checked in the induced graph. Recently, CNDPs on special classes of graphs including split graphs, bipartite graphs and graphs of bounded treewidth were considered in [Addis *et al.* (2013)]. Lalou *et al.* (2016) proposed a new variant of this problem, called the component-cardinality-constrained critical node problem (3C-CNP). This variant seeks to find a minimal set of nodes, the removal of which constrains the size of each connected component in the induced graph to a given bound. Aringhieri *et al.* (2018) studied the distance critical node problem. It is a generalization of the critical node problem where the distances between node pairs impact on the objective function. Dinh and Thai (2013) and Dinh *et al.* (2010) presented a new formulation of CNDP called the β-vertex separator problem. They studied the complexity and inapproximability on general graphs and proposed a pseudo-approximation method and a heuristic approach to solve this problem on general graphs. Similarly, Shen and colleagues (Shen *et al.*, 2013a,b) provided complexity analysis for CNDPs on general graphs and power-law graphs. Aringhieri *et al.* (2016a) and Veremyev *et al.* (2014b) used the above-mentioned three different connectivity measures, took into account both the budget and connectivity constraints, and obtained six different variants of the CNDP.

There are exact and heuristic algorithms proposed for the CNDP in the literature. The exact algorithms for the CNDP include an Integer Linear Program (ILP) for problems with a potentially non-polynomial number of constraints [Di Summa *et al.* (2012)]. Other models with a polynomial number of constraints were also studied in [Veremyev *et al.* (2014b)] and [Veremyev *et al.* (2014a)]. A recent work by Pavlikov (2018) provided a model with a polynomial number of constraints, which has the same linear relaxation as the model of [Di Summa *et al.* (2012)]. Addis *et al.* (2013) defined a dynamic programming recursion that solves the problem in polynomial time when the graph has bounded treewidth and unit connection costs. Besides, an approximation algorithm named the bi-criteria randomized rounding approach was proposed in [Ventresca and Aleman (2014)]. However, these exact methods offer limited applicability since most of them are based on the ILP formulation. Recently, several heuristic algorithms have been proposed for the CNDP, for example, multiple greedy constructive heuristics [Addis *et al.* (2016); Ventresca and Aleman (2015)] and local search metaheuristics [Aringhieri *et al.* (2016b); Ventresca and Aleman (2014)]. A simulated annealing algorithm and a population based incremental learning algorithm without approximation bounds were applied to the CNDP with up to 5000 nodes [Ventresca (2012)]. A fast greedy algorithm has been recently presented for approximating solutions for large-scale networks [Ventresca and Aleman (2014)]. A variable neighborhood search that outperforms the population-based method [Ventresca (2012)] was proposed in [Aringhieri *et al.* (2015)]. Purevsuren *et al.* (2017) provided results competitive with those of [Aringhieri *et al.* (2016a)] for the single-objective CNDP. Readers can refer to a recent survey by Lalou *et al.* (2018) for a detailed exposition of the CNDP and relevant results in the literature.

Most of the aforementioned formulations and algorithms regard the CNDP as single-objective problems. These formulations assume that decision makers either have prior knowledge on the maximum number of nodes that can be removed or have the ability to give a threshold of the induced network connectivity. However, it is not easy for decision makers to gain preference knowledge since they know little about the problem itself, and the CNDP is not an exception. In fact, the connectivity of an induced network and the cost of removing nodes are two conflicting objectives and should be considered simultaneously. Aringhieri *et al.* (2016a) are the first authors to propose fully bi-objective results for the CNDP and display PFs for some instances. Furthermore, Faramondi *et al.* (2018) also considered the CNDP as a bi-objective problem and firstly provided an explicitly bi-objective approach, i.e., a multi-objective ant colony optimization algorithm. It should be noted that the CNDP is also recognized as a bi-objective problem in [Ventresca *et al.* (2018)], but in a different way. To be specific, Ventresca *et al.* (2018) tackled a bi-objective CNDP where the two objectives are the number of components and the variance of their cardinality, which is like a generalization of pairwise connectivity. This chapter studies a bi-objective formulation of the CNDP (Bi-CNDP), which considers both the connectivity of an induced network and the cost of removing nodes and presents related theoretical and computational results.

For weighted networks, we assume that each node is assigned with a weighted value that is related to the cost of removing it, and decision makers want to minimize the pairwise connectivity of an induced graph and minimize the cost of removing these nodes at the same time. We first prove the NP-hardness on general graphs and the existence of a polynomial algorithm for constructing an ε-approximated PF for CNDPs on trees. Then different types of mating pools and replacement pools are proposed and embedded in decomposition-based MOEAs. Two state-of-the-art decomposition-based MOEAs including the MOEA/D and DMOEA-εC are modified and applied to solve the Bi-CNDP. Numerical results on 16 modified famous benchmark problems are conducted to assess the effectiveness of different mating pools and replacement pools. Further computational results demonstrate different performances of two improved MOEAs, i.e., I-MOEA/D and I-DMOEA-εC, on solving the Bi-CNDP. Finally, a decision making process from the perspective of minimizing the pairwise connectivity of the induced graph given a constraint on the cost of removing nodes is proposed for helping decision makers to identify the most critical nodes.

6.2   Mathematical Formulations and Theoretical Results of Bi-CNDP

This section first recalls some basic definitions related to the CNDP and then presents mathematical formulations of the Bi-CNDP. Next, some important theoretical results related to the complexity analysis of the Bi-CNDP are illustrated.

6.2.1   *Mathematical Formulations of Bi-CNDP*

Let $G = (V, E, C)$ be a weighted and undirected graph composed of n nodes $V = \{v_1, \ldots, v_n\}$ with weight values related to the cost of removing these nodes $C = \{c_1, \ldots, c_n\}$ and m edges $E = \{(v_i, v_j), i, j = 1, \ldots, n\}$, where $(v_i, v_j) \in E \subseteq V \times V$ captures the existence of a relation between node v_i and node v_j. First, some basic definitions of graphs that will be used in the following are recalled.

Definition 6.1 (Path). A path over an undirected graph $G = (V, E, C)$ starting at a node $v_i \in V$ and ending at a node $v_j \in V$ is a subset of links in E that connects nodes v_i and v_j.

Definition 6.2 (Connected). An undirected graph G is connected if for each pair of nodes v_i and v_j there is a path over G that connects them.

Definition 6.3 (Connected Component). A connected component of G is a connected subgraph $G_i = (V_i, E_i)$ such that, over G_i (i.e., $V_i \subseteq V, E_i \subseteq E$), no node in V_i is connected to a node in $V \backslash V_i$.

Here, the pairwise connectivity (PWC) [Arulselvan *et al.* (2009)] is adopted as a measure of connectivity of a graph. The PWC represents the number of distinct node pairs connected by a path over G. To be specific, the definition of $PWC(G)$ is given as follows:

$$PWC(G) = \frac{1}{2} \sum_{v_i, v_j \in V, v_i \neq v_j} x_{ij}, \tag{6.1}$$

where x_{ij} equals 1 if node v_i and node v_j are connected via a path in G, and it is equal to 0 otherwise. $PWC(G)$ is monotonically non-increasing with respect to edge removals, since the removal of an edge cannot increase the number of pairs of nodes connected by a path. Moreover, $PWC(G) = \frac{n(n-1)}{2}$ holds for a graph G when all pairs of nodes are connected. Therefore, we define the normalized pairwise connectivity as:

$$nPWC(G) = \frac{1}{n(n-1)} \sum_{v_i,v_j \in V, v_i \neq v_j} x_{ij}. \tag{6.2}$$

It is obvious that $nPWC(G) \in [0,1]$. It is straightforward to note that $nPWC(G)$ can be regarded as a measure of the degree of connectivity of G, since $nPWC(G)$ is proportional to the fraction of node pairs that are connected at least via one path. The larger $nPWC(G)$ is, the closer G is to a connected graph.

The first goal of the Bi-CNDP is to determine a subset of nodes $R \subseteq V$ such that the induced residual graph $G(V \backslash R)$ has minimum pairwise connectivity $nPWC(G(V \backslash R))$. Apart from minimizing the degree of connectivity of the reduced network after the removal, decision makers also consider minimizing the cost of removing selected nodes. Specifically, the cost of removing selected nodes $R \subseteq V$ from G is given as:

$$nCost(R) = \sum_{v_i \in R} c_i \Big/ \sum_{v_i \in G} c_i, \tag{6.3}$$

where c_i represents the weight value associated with node $v_i \in V$. $\sum_{v_i \in G} c_i$ is introduced so that the two objectives are comparable, as they both have values in $[0,1]$. In conclusion, the formulation of Bi-CNDP is presented as:

$$
\begin{aligned}
\text{minimize} \quad & (nPWC(G(V \backslash R)), nCost(R)) \\
& x_{ij} + y_i + y_j \geq 1, \forall v_i, v_j \in V. \\
\text{s.t.} \quad & x_{ij} + x_{jk} + x_{ki} \neq 2, \forall v_i, v_j, v_k \in V \\
& x_{ij} \in \{0,1\}, y_i \in \{0,1\}, \forall v_i, v_j \in V,
\end{aligned}
\tag{6.4}
$$

where $\mathbf{y} = (y_1, y_2, \ldots, y_n)$ is the decision vector, y_i is a boolean variable, $y_i = 1$ if node $v_i \in V$ is removed from the original graph and $y_i = 0$ otherwise. x_{ij} represents the connectivity between node v_i and v_j, and it equals to 1 if node v_i and node v_j are in the same connected component of $G(V \backslash R)$; otherwise it equals to 0.

The first constraint enforces the separation of nodes to different components. To be specific, deleted nodes do not share an edge to any other node by setting edge $x_{ij} = 0$ as deleted if either or both nodes v_i and v_j are deleted. The second constraint is concerned with a triangle inequality. That is to say, if nodes v_i and v_j, and v_j and v_k are in the same component, then nodes v_i and v_k must be in the same component. In this formulation, there is no need to specify a hierarchy between two objectives nor to gain prior information about the psychology of decision makers.

6.2.2 *Theoretical Results on Bi-CNDP*

Firstly, we transform the Bi-CNDP into a single-objective formulation (denoted as the S-CNDP), as shown in (6.5) by converting the first objective function into an additional constraint function. Then, we can get the following results.

$$
\begin{aligned}
\text{minimize} \quad & nPWC(G(V \backslash R)) \\
& nCost(R) \leq C, \forall v_i, v_j \in V \\
s.t. \quad & x_{ij} + y_i + y_j \geq 1, \forall v_i, v_j \in V \\
& x_{ij} + x_{jk} + x_{ki} \neq 2, \forall v_i, v_j, v_k \in V \\
& x_{ij} \in \{0, 1\}, y_i \in \{0, 1\}, \forall v_i, v_j \in V.
\end{aligned}
\tag{6.5}
$$

Theorem 6.1. *The Bi-CNDP is strongly NP-hard on general graphs.*

The proof of Theorem 6.1 is straightforward. Since the S-CNDP formulated in (6.5) is just the K-CNP described in [Arulselvan *et al.* (2009)], the S-CNDP is strongly NP-complete on general graphs [Addis *et al.* (2013)]. Furthermore, we know on general grounds that when the complexity status of an ε-constrained version of a multi-objective problem (MOP) is demonstrated, the MOP can only be harder to solve, which implies that the Bi-CNDP is strongly NP-hard. Indeed if it was not, a solution for each ε-constrained version could be obtained in (pseudo-)polynomial time, which would imply that the S-CNDP is not NP-hard. This leads to a contradiction. Hence, the Bi-CNDP is strongly NP-hard on general graphs.

The subclass of the CNDP over trees was studied in [Di Summa *et al.* (2011)]. It has been proved that the CNDP over trees is still NP-complete when general connection costs are specified. However, the cases where all connections have unit cost are solvable in polynomial time by dynamic programming approaches. Besides, a dynamic programming recursion that solves the problem in polynomial time when the graph has bounded treewidth was proposed in [Addis *et al.* (2013)]. Besides, the complexity of removing larger node structures has also received significant attention. Granata and colleagues introduced the concept of a critical disruption path, which refers to a path between a source and a destination vertex whose deletion minimizes the cardinality of the largest remaining connected component. The proposed network interdiction model seeks to optimally disrupt network operations. In [Walteros *et al.* (2018)], the authors have proved that the problems of removing critical cliques, stars and connected subgraphs are all strongly NP-complete.

Finding all Pareto optimal solutions is often computationally problematic for multi-objective discrete optimization, since there are usually exponentially (or infinite) large Pareto optimal solutions. Furthermore, for the simplest problems with two objectives, determining whether a point belongs to the Pareto optimal set is NP-hard [Papadimitriou and Yannakakis (2000)]. One way to handle these problems is to introduce the ε-approximated Pareto set $P_\varepsilon(x)$ whose definition is given as follows.

Definition 6.4 (ε-approximated Pareto Set $P_\varepsilon(x)$). Given a scalar $\varepsilon > 0$, an ε-approximate Pareto optimal set, denoted by $P_\varepsilon(x)$, is a subset of X such that there is no other solution $y \in X$ such that $(1+\varepsilon) \cdot f_i(y) \le f_i(x)$ for all $x \in P_\varepsilon(x)$ and for some $i \in \{1, 2, \ldots, m\}$, where X is a nonempty feasible set for a certain MOP and m represents the number of objectives of the MOP.

This definition says that every other solution is almost dominated by some solution in $P_\varepsilon(x)$, i.e., there is a solution in $P_\varepsilon(x)$ that is within a factor of ε in all objectives. According to theoretical results stated in [Papadimitriou and Yannakakis (2000)] and the fact that the CNDP is polynomially solvable on trees via dynamic programming [Di Summa *et al.* (2011)], we can have the following statement.

Theorem 6.2. *For CNDPs on trees, there is a polynomial algorithm in n and $1/\varepsilon$ for constructing the approximate Pareto curve $P_\varepsilon(x)$ for the Bi-CNDP formulated in (6.4), where n represents the size of an instance of the Bi-CNDP.*

However, there is no such property for Bi-CNDPs on general graphs. Thus, the exponential size of feasible solutions for Bi-CNDPs on general graphs calls for the use of heuristic algorithms to find Pareto optimal solutions for the proposed Bi-CNDP.

6.3   Improved Decomposition-based Multi-objective Evolutionary Algorithms

Among various heuristic algorithms, the decomposition-based MOEAs have gained much attention during these decades. This chapter modifies two decomposition-based MOEAs to deal with the Bi-CNDP. Thus this section presents a brief description of two decomposition-based MOEAs and their variants, which will all be used in the next section.

Decomposition is an efficient and prevailing strategy for solving MOPs. In decomposition-based MOEAs, an MOP is decomposed into a number of scalar subproblems by using various scalarizing functions. The weighting method, Tchebycheff approach, boundary intersection method, and the ε-constraint method are classical generation methods in the field of mathematical programming and have been adopted for the multi-objective optimization. Zhang and colleagues adopted the first three aggregation functions and proposed the multi-objective evolutionary algorithm based on decomposition (MOEA/D). Many variants such as MOEA/D-DRA [Zhang *et al.* (2009a)], MOEA/D-AWA [Qi *et al.* (2014)], and so on, have been investigated.

Recently, a new MOEA named the decomposition-based multi-objective evolutionary algorithm with the ε-constraint framework (DMOEA-εC) was proposed in [Chen *et al.* (2017)] and has demonstrated its superiority over a number of MOEAs. DMOEA-εC firstly incorporates the ε-constraint method into the decomposition strategy and decomposes an MOP into a series of scalar-constrained optimization subproblems by assigning each subproblem with an upper bound vector. In decomposition-based MOEAs, all subproblems are optimized simultaneously by only using information from neighboring subproblems. The decomposition-based MOEAs have lower computational complexity than Pareto-based and indicator-based MOEAs. Besides, an external archive population EP is added to store non-dominated solutions found so far. Details of MOEA/D and DMOEA-εC can be found in [Li and Zhang (2009)] and [Chen *et al.* (2017)], respectively.

Performances of these two MOEAs have been witnessed on continuous and discrete benchmark problems, but not on the Bi-CNDP. Thus, it is reasonable to adopt them to solve the proposed Bi-CNDP. When applying two MOEAs to the Bi-CNDP, there are two important issues that need to be discussed: 1) How to determine the mating pool for selecting parent candidates? 2) How to determine the replacement pool to replace old candidates after a new candidate is generated? The above-mentioned two issues will be discussed in the next two subsections.

6.3.1   *The Mating Pool*

Given a certain recombination operator, the mating pool plays a vital role in generating new solutions. In this chapter, for certain solutions of a subproblem, we discuss four types of mating pools:

1) -N: The set of solutions of neighboring subproblem of a subproblem serves as the mating pool, and parent solutions are randomly selected from the mating pool.
2) -P: The whole population is regarded as the mating pool, and parent solutions are randomly selected from the whole population.
3) -NP: The parent solutions are selected from the set of solutions of neighboring subproblem of a subproblem with a probability δ. Parent solutions are selected from the whole population with a probability $1 - \delta$.
4) -NP-EP: One of the parent solutions is selected randomly from the external archive population EP, and the other one is selected according to the third approach.

The four ways of determining the mating pool will be added into the MOEA/D, thus obtaining four variants including MOEA/D-N, MOEA/D-P, MOEA/D-NP and MOEA/D-NP-EP. Similary, four variants of DMOEA-εC, i.e., DMOEA-εC-N, DMOEA-εC-P, DMOEA-εC-NP and DMOEA-εC-NP-EP can be developed.

6.3.2 The Replacement Pool

When a new candidate is generated, it will be used to compare with and update other old candidates in the replacement pool. For a new solution of one subproblem, we investigate effects of two types of replacement pools in decomposition-based MOEAs.

1) -L: A newly generated candidate is compared with the set of solutions of neighboring subproblems of a subproblem, thus the replacement takes place locally.
2) -G: The whole population serves as the replacement pool, which implies a global replacement strategy.

Two types of replacement pool will be integrated into MOEA/D, thus obtaining its variants, i.e., MOEA/D-L and MOEA/D-G. Similarly, two variants of DMOEA-εC, including DMOEA-εC-L and DMOEA-εC-G, are designed.

6.4 Experimental Design

This section is devoted to experimental design for demonstrating the overall quality of solutions found by two MOEAs and their variants over 4 sets of 16 benchmark instances proposed in [Ventresca (2012)]. First, details of the 16 benchmark problems are outlined, then parameter settings are provided. Finally, the experimental results are illustrated. We compare results obtained via two MOEAs and their variants to evaluate the performance of different strategies of determining the mating pool and the replacement pool. Further experiments are conducted to demonstrate the different performances of two improved MOEAs on solving the Bi-CNDP. Additionally, a decision-making process based on obtained non-dominated solutions is illustrated.

6.4.1 *Benchmark Problems*

The benchmark set is composed of the graphs proposed in [Ventresca (2012)], and many results are available for these graphs as single-objective problems. This data set contains 16 undirected, unweighted graphs belonging to 4 groups that are created by complex network generator algorithms. Barabasi-Albert (BA) graphs are scale-free networks and proved to be the easiest to process. While the Watts–Strogatz (WS) graphs are designed to mimic a small-world structure with a denser structure, they turn out to be the most challenging ones. Erdos–Renyi (ER) graphs are random graphs and Forest–Fire (FF) graphs reproduce the behavior of how a fire spreads through a forest. None of these graphs is expected to reproduce a real network. However, real networks usually display a mixture of these characteristics. Further information about these networks can be found in [Ventresca (2012)].[1]

In order to characterize these graphs precisely, Table 6.1 displays the following quantities: the number of nodes n, the number of edges m, the average degree $< d >= 2 \cdot m/n$, the number of articulation points nAP,[2] the value of the clustering coefficient CC, the average shortest path length D [Aringhieri *et al.* (2016a)], the number of nodes having degree 1 $|D_1|$, and the number of nodes which are neighbors of those in D_1 [Veremyev *et al.* (2014a)], denoted as $|N(D_1)|$ [Aringhieri *et al.* (2016a)]. The number of articulation points (nAP) is taken into account since a larger fraction of

---

[1]The numbers of nodes and edges of the ER, WS and FF graphs in the main body of [Ventresca (2012)] are slightly different from the data set obtained from the website given in [Ventresca (2012)]. Here, these characteristics are given based on the data set downloaded from the website.

[2]A vertex in an undirected connected graph is an articulation point if and only if removing it disconnects the graph. Articulation points represent vulnerabilities in a connected network.

Table 6.1 Main Characteristics of 16 Benchmark Instances.

Instance	n	m	$\langle d \rangle$	nAP	CC	D	$|D_1|$	$N(D_1)$
BA500	500	499	1.996	164	0.000	5.663	336	149
BA1000	1000	999	1.998	324	0.000	6.045	676	290
BA2500	2500	2499	1.999	825	0.000	6.901	1675	729
BA5000	5000	4999	1.999	1672	0.000	8.380	3328	1475
WS250	250	1246	9.968	0	0.473	3.327	0	0
WS500	500	1496	5.984	0	0.420	5.304	0	0
WS1000	1000	4996	9.992	0	0.483	4.444	0	0
WS1500	1500	4498	5.997	0	0.480	7.554	0	0
ER235	235	350	2.979	48	0.006	5.339	39	37
ER466	466	700	3.004	84	0.002	5.974	69	64
ER941	941	1400	2.976	177	0.005	6.559	147	139
ER2344	2344	3500	2.986	419	0.001	7.516	396	354
FF250	250	514	4.112	83	0.276	4.816	57	50
FF500	500	828	3.312	195	0.247	6.026	160	136
FF1000	1000	1817	3.634	362	0.216	6.173	280	236
FF2000	2000	3413	3.413	725	0.245	7.587	552	477

articulation points usually results in a graph that is easier to fragment. The clustering coefficient CC signals the tendency of nodes to be clustered together. The average shortest path length D indicates the average distance between two nodes taken at random inside the graph.

Since these networks are unweighted in the original, new benchmark instances are created by assigning a weight value to each node of each network. The weight value of each node is regarded as the cost of removing it. We adopt the weight generation method used in [Ventresca *et al.* (2018)], as described in the following:

1) Weights are randomly assigned, where $c(v) \in [0.2, 3]$, $\forall v \in V$;
2) Weights are logarithmicly assigned with node degree d_v, where $c(v) = \log(d_v) + 0.5$, $\forall v \in V$.

6.4.2 Parameter Settings

For fair comparison, the choice of parameters remains the same for two MOEAs. Specifically, we adopt binary vectors as encoding schemes for solutions. The population size N is set to 300, 400, 500 and 600 for benchmark problems whose number of nodes is $n \leq 500$, $500 < n \leq 1000$, $1000 < n \leq 2500$, and $2500 < n \leq 5000$, respectively. For fair comparison, an external population with the size of $S = \lfloor 1.5 \cdot N \rfloor$ is added to each

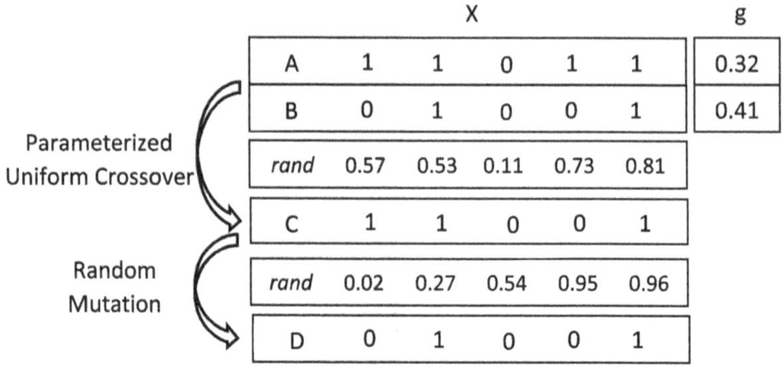

Fig. 6.1   An example of the parameterized uniform crossover and the random mutation where the number of nodes is 5. A and B are parent solutions. C and D are offspring solutions after crossover and mutation, respectively. *rand* denotes a uniformly randomly distributed value in $[0, 1]$ and g represents the objective value of each solution (i.e., subproblem).

algorithm, where $\lfloor \cdot \rfloor$ returns the nearest integer in the direction of negative infinity.

Besides, the parameterized uniform crossover [Spears and Jong (1991)] and random mutation [Arulselvan *et al.* (2011)] are adopted in generating new solutions. Moreover, control parameters for these reproduction operators are the same as those used in [Arulselvan *et al.* (2011)]. To be specific, the biased probability of crossover is set as 0.65 and the random mutation probability for each variable of a solution is set as 0.03. Figure 6.1 provides an example of the parameterized uniform crossover and the random mutation where the number of nodes is 5.

For MOEA/D, DMOEA-εC and their variants, the neighborhood size is $T = \lfloor 0.1 \cdot N \rfloor$, the probability of selecting mate solutions from neighborhood is $\delta = 0.9$, and the maximal number of replacement is $n_r = \lfloor 0.01 \cdot N \rfloor$. Inspired by [Ventresca (2012)], the maximum number of iterations I is set as 2500, 4000, 6000 and 7500 for test instances whose number of nodes n is $n \leq 500$, $500 < n \leq 1000$, $1000 < n \leq 2500$, and $2500 < n \leq 5000$, respectively. For DMOEA-εC and its variants, the iteration interval of alternating the main objective function IN_m is set to $\lfloor 20\% \cdot I \rfloor$. Both algorithms stop when the number of iterations reaches the maximum number, and each algorithm is executed 20 times independently on each instance.

6.4.3 *Experimental Results*

This section includes two parts. The first part compares the performance of different variants of two decomposition-based MOEAs, and the second part compares the performance of two identified MOEAs obtained according to the experimental results of the first part.

Two commonly used performance metrics, i.e., inverted generational distance (IGD) [Zhou *et al.* (2005)] and hypervolume (HV) [Zitzler and Thiele (1999)] are employed to evaluate the performance of compared algorithms.

In calculating performance metrics, N non-dominated solutions are selected from the external population using the crowding distance approach. With the purpose of calculating the IGD metric value, P^* is chosen to be the set of non-dominated solutions extracted from the combination of all solutions obtained via two MOEAs and their various variants over 20 independent runs. In order to compute the HV metric value, the reference point is set as 1.1 times the estimated nadir point[3] based on P^* for each instance.

6.4.3.1 *Comparison among various variants*

As mentioned above, the determination of the mating pool and the replacement pool are important for evolving the population towards the desired PF. Therefore, effects of various variants of two MOEAs with different strategies of determining the mating pool and the replacement pool are deeply analyzed on all test problems with random and logarithmic weights. The means and standard deviations of IGD metric values over 20 runs of MOEA/D and its variants about the mating pool determination on all test instances with random weights are shown in Table 6.2. Additionally, the means and standard deviations of HV metrics over 20 runs of variants of DMOEA-εC-NP-EP about the replacement pool determination on all test instances with logarithmic weights are shown in Table 6.3. The bold data in each table are the best mean metric values for each instance. The mean IGD (HV) values for each instance are sorted in an ascending (descending) order, and the numbers in the square brackets are their ranks. Besides, the mean rank values in terms of IGD and HV metrics over all test instances are displayed for each variant to have a global view of the performance of all variants.

---

[3]The nadir point is the upper bound of the PF.

Table 6.2 Statistical Results (Mean[Rank](Std.)) of MOEA/D and its Variants with Different Types of Mating Pools over 20 Independent Runs on the 16 Instances with Random Weights in Terms of *IGD* Metrics.

Instance	MOEA/D-N	-P	-NP	-NP-EP
BA500	3.80E-03[3](1.57E-04)	3.86E-03[4](1.70E-04)	3.51E-02[2](2.26E-04)	**3.31E-03[1](4.60E-04)**
BA1000	2.58E-02[3](1.05E-03)	2.97E-02[4](8.12E-03)	1.74E-02[2](1.92E-03)	**1.50E-02[1](3.28E-03)**
BA2500	5.63E-04[3](1.93E-04)	5.77E-04[4](1.25E-04)	5.36E-04[2](1.06E-04)	**5.24E-04[1](7.50E-04)**
BA5000	8.64E-03[3](7.33E-04)	8.82E-03[4](6.29E-04)	8.59E-03[2](3.09E-04)	**8.29E-03[1](7.35E-04)**
WS250	3.53E-02[3](6.87E-03)	3.68E-02[4](2.63E-03)	**2.94E-02[1](1.47E-03)**	3.15E-02[2](1.29E-03)
WS500	1.88E-02[3](1.40E-03)	1.96E-03[4](4.59E-03)	**1.45E-02[1](3.45E-03)**	1.47E-02[2](4.78E-03)
WS1000	2.16E-02[3](1.14E-03)	2.24E-02[4](1.85E-03)	1.77E-02[2](1.62E-03)	**1.71E-02[1](1.47E-03)**
WS1500	5.54E-02[3](3.40E-03)	5.60E-02[4](6.06E-03)	**5.20E-02[1](4.12E-03)**	5.42E-02[2](3.50E-03)
ER235	3.75E-02[2](1.08E-03)	4.07E-02[4](2.62E-03)	3.96E-02[3](3.75E-03)	**3.55E-02[1](2.13E-03)**
ER466	9.82E-02[4](5.60E-03)	9.70E-02[3](6.49E-03)	9.26E-02[2](5.12E-03)	**9.22E-03[1](7.89E-03)**
ER941	3.56E-02[3](2.01E-03)	3.73E-02[4](2.81E-03)	3.33E-02[2](1.11E-03)	**3.14E-02[1](1.04E-03)**
ER2344	3.19E-02[3](2.09E-03)	3.26E-02[4]1.99E-03)	**2.88E-02[1](1.50E-03)**	2.92E-02[2](1.63E-03)
FF250	4.79E-02[3](4.71E-03)	5.41E-02[4](2.91E-03)	4.71E-02[2](4.02E-03)	**4.63E-02[1](3.04E-03)**
FF500	4.38E-02[2](3.03E-03)	5.08E-02[4](1.64E-03)	4.76E-02[3](1.08E-03)	**4.29E-02[1](1.30E-03)**
FF1000	3.31E-02[3](1.73E-03)	3.83E-01[4](1.31E-03)	2.75E-02[2](1.97E-03)	**2.61E-02[1](1.08E-03)**
FF2000	2.89E-02[4](1.05E-03)	2.88E-02[3](6.28E-03)	**2.54E-02[1](2.40E-03)**	2.60E-02[2](1.89E-03)
Rank	3.000	3.875	1.813	1.313

Table 6.3   Statistical Results (Mean[Rank](Std.)) of DMOEA-εC-NP-EP and its Variants with Different Types of Replacement Pools over 20 Independent Runs on the 16 Instances with Logarithmic Weights in Terms of HV Metrics.

Instance	DMOEA-εC-NP-EP-L	-G
BA500	1.71E-01[2](1.41E-02)	**1.89E-01[1](1.77E-02)**
BA1000	5.58E-04[2](3.45E-05)	**5.95E-04[1](1.70E-05)**
BA2500	**4.68E-05[1](1.17E-05)**	4.54E-05[2](1.01E-05)
BA5000	6.37E-05[2](1.78E-05)	**6.58E-05[1](2.08E-05)**
WS250	4.16E-01[2](6.41E-02)	**4.49E-01[1](2.97E-02)**
WS500	2.46E-01[2](4.03E-02)	**2.77E-01[1](3.71E-02)**
WS1000	6.11E-02[2](5.67E-03)	**6.51E-02[1](2.36E-03)**
WS1500	7.10E-02[2](5.14E-03)	**7.73E-02[1](3.93E-03)**
ER235	5.37E-02[2](2.36E-03)	**5.78E-02[1](1.44E-03)**
ER466	2.01E-01[2](1.62E-02)	**2.16E-01[1](2.73E-02)**
ER941	**1.73E-02[1](5.03E-03)**	1.68E-02[2](4.89E-03)
ER2344	**6.23E-04[1](9.39E-05)**	6.01E-04[2](8.81E-05)
FF250	5.25E-01[2](2.30E-02)	**5.62E-01[1](3.84E-02)**
FF500	**6.21E-02[1](1.41E-03)**	6.09E-02[2](2.77E-03)
FF1000	7.35E-03[2](7.04E-03)	**7.74E-03[1](4.58E-03)**
FF2000	9.43E-03[2](8.01E-04)	**9.89E-03[1](7.23E-04)**
Rank	1.750	1.250

As can be seen in Table 6.2, in terms of *IGD* metric values, MOEA/D-NP-EP shows obvious advantage over other variants on the majority of test instances with random weights. On the other instances, MOEA/D-NP performs best. Table 6.2 also reveals the overall rank of the four variants, that is, MOEA/D-NP-EP, MOEA/D-NP, MOEA/D-N and MOEA/D-P according to the mean rank values. Results in Table 6.2 highlight the effectiveness of the fourth strategy of determining the mating pool. To be specific, a parent solution is first selected from the external archive populution EP. Then, the other one is selected from the neigborhood with a probability δ, and it is selected from the whole population with a probability $1 - \delta$. As to DMOEA-εC and its variants about the mating pool determination, similar results can be obtained. That is, DMOEA-εC with the fourth type of mating pool performs the best among all variants.

The superiority of the fourth mating pool determination strategy can be explained in the following two aspects. Firstly, the basic assumption of decomposition-based MOEAs is that neighboring subproblems have similar optimal solutions. Thus, two solutions of neighboring subproblems have a higher chance to produce good solutions, which can accelerate convergence.

The participation of the whole population in the mating pool with a probability 1-δ is beneficial to give birth to diverse offsprings. Additionally, since
the external archive population EP is used to store non-dominated solutions found so far, it is reasonable to make best use of it to produce new
solutions with high quality.

It can be observed from Table 6.3 that DMOEA-εC-NP-EP-G outperforms DMOEA-εC-NP-EP-L on the majority of test instances with logarithmic weights in terms of the HV metrics. Similar results can be obtained
for MOEA/D and its variants about the replacement pool determination
in terms of the HV metrics. The broadened range of the replacement
pool makes the replacement more effective, which is good for convergence.
Besides, the limited number of replacement takes control of maintaining the
diversity of a population. In conclusion, the effectiveness of the fourth strategy of determining the mating pool and the global replacement strategy
are confirmed experimentally. Thus, the improved MOEA/D and DMOEA-
εC with the fourth type of mating pool and the global replacement pool
are denoted as I-MOEA/D and I-DMOEA-εC, respectively. They will be
employed in the following numerical experiments.

6.4.3.2   *Comparison between two improved MOEAs: I-MOEA/D and I-DMOEA-εC*

This part of the experiment is designed to study the effectiveness of I-
MOEA/D and I-DMOEA-εC on Bi-CNDPs. Our comparison is made of
two perspectives: 1) the comparison between single-objective and multi-
objective formulations, and 2) the comparison between two improved multi-
objective approaches.

Firstly, it has been theoretically proven that the solution of a single-
objective problem whose objective is a convex linear combination of the
objectives of the MOP is part of the PF of the MOP [Miettinen (1999)].
This statement is still valid for the Bi-CNDPs, and the optimization of the
Bi-CNDP can present a set of Pareto optimal solutions for decision makers
to have a global view of the problem and make more reasonable decisions.

Secondly, concerning the latter perspective, the IGD [Zhou *et al.* (2005)]
and HV [Zitzler and Thiele (1999)] are still employed to evaluate the performance of compared algorithms. The means and standard deviations of
IGD and HV metric values of two improved MOEAs over 20 independent
runs of each algorithm on 16 instances with random and logarithmic weights
are shown in Tables 6.4 and 6.5. The Wilcoxon's rank sum test at a 95%
significance level is conducted to test the significance of differences between

Table 6.4 Statistical Results (Mean(Std.)) of 2 Improved MOEAs over 20 Independent Runs on the 16 Instances with Random and Logarithmic Weights in Terms of *IGD* Metrics.

Instance	Random Weights		Logarithmic Weights	
	I-MOEA/D	I-DMOEA-εC	I-MOEA/D	I-DMOEA-εC
BA500	3.19E-03(1.79E-04)	**1.74E-04**†(5.29E-04)	3.41E-02(1.39E-03)	**2.13E-04**†(1.26E-05)
BA1000	1.57E-02(7.84E-03)	**5.76E-04**†(6.01E-05)	6.16E-03(6.88E-04)	**4.21E-04**†(1.01E-05)
BA2500	4.74E-04(2.21E-05)	**4.64E-04**≈(3.02E-05)	2.77E-02(1.86E-03)	**6.19E-04**†(3.72E-05)
BA5000	7.57E-03(8.13E-04)	**7.34E-04**†(1.30E-05)	6.05E-03(5.38E-04)	**5.48E-04**†(5.53E-05)
WS250	2.23E-02(1.02E-03)	**9.36E-04**†(1.90E-05)	2.28E-02(1.40E-03)	**4.82E-04**†(3.47E-05)
WS500	1.50E-02(1.39E-03)	**6.23E-04**†(3.78E-05)	7.86E-03(2.66E-03)	**6.16E-03**≈(5.53E-04)
WS1000	1.65E-02(1.72E-03)	**1.40E-03**†(6.18E-04)	9.84E-03(8.79E-04)	**8.15E-04**†(6.76E-05)
WS1500	4.89E-02(2.72E-03)	**4.27E-03**†(2.18E-04)	7.15-03(4.77E-04)	**6.42E-04**†(4.61E-05)
ER235	3.52E-02(1.89E-03)	**2.64E-04**†(3.47E-05)	2.19E-02(5.90E-03)	**5.92E-04**†(7.76E-05)
ER466	8.79E-03(8.40E-04)	**8.64E-04**†(5.81E-05)	1.42E-02(1.12E-03)	**2.62E-03**†(1.40E-04)
ER941	2.70E-02(2.29E-03)	**4.62E-03**†(5.73E-04)	2.35E-03(2.91E-04)	**2.52E-03**≈(1.63E-04)
ER2344	2.28E-02(3.79E-03)	**6.99E-03**†(5.31E-04)	7.54E-03(8.93E-04)	**6.17E-04**†(6.04E-05)
FF250	5.20E-02(2.24E-03)	**9.84E-04**†(6.48E-05)	5.19E-02(5.23E-03)	**4.10E-03**†(2.55E-04)
FF500	3.75E-02(4.71E-03)	**2.73E-03**†(7.68E-04)	1.30E-02(1.43E-03)	**9.19E-04**†(1.30E-05)
FF1000	2.28E-02(7.25E-03)	**1.33E-03**†(4.03E-04)	9.05E-02(6.20E-03)	**6.81E-03**†(1.64E-04)
FF2000	2.58E-02(3.50E-03)	**2.81E-03**†(1.92E-04)	3.24E-02(4.18E-03)	**4.46E-03**†(5.11E-04)

Table 6.5 Statistical Results (Mean(Std.)) of 2 Improved MOEAs over 20 Independent Runs on the 16 Instances with Random and Logarithmic Weights in Terms of HV Metrics.

Instance	I-MOEA/D Random Weights	I-DMOEA-εC Random Weights	I-MOEA/D Logarithmic Weights	I-DMOEA-εC Logarithmic Weights
BA500	9.15E-03(3.78E-04)	**6.81E-02†(2.12E-03)**	5.25E-02(3.31E-03)	**1.89E-01†(2.71E-02)**
BA1000	8.81E-05(1.13E-05)	**4.33E-04†(2.29E-05)**	1.39E-04(1.67E-05)	**5.95E-04†(6.61E-05)**
BA2500	8.87E-06(8.90E-06)	4.55E-05†(3.87E-06)	**5.59E-05(6.57E-06)**	4.54E-05§(7.12E-06)
BA5000	6.48E-05(8.76E-05)	**7.56E-05†(6.45E-06)**	4.99E-05(1.50E-06)	**6.58E-05≈(4.17E-06)**
WS250	3.44E-01(2.29E-02)	**4.04E-01†(1.86E-02)**	3.41E-01(2.41E-02)	**4.49E-01†(1.79E-02)**
WS500	2.89E-01(3.77E-02)	**3.57E-01†(2.91E-02)**	1.58E-01(3.32E-02)	**2.77E-01†(1.51E-02)**
WS1000	**3.95E-02§(5.69E-03)**	3.17E-02§(7.54E-03)	**1.75E-01(3.74E-02)**	6.51E-02§(1.38E-03)
WS1500	6.61E-02†(7.17E-03)	**7.67E-02†(9.54E-04)**	2.54E-02(3.12E-03)	**7.73E-02†(4.91E-03)**
ER235	2.98E-01(1.97E-02)	**4.23E-01†(5.16E-02)**	3.05E-01(4.18E-02)	**5.78E-01†(4.63E-02)**
ER466	1.16E-01(1.53E-02)	**3.45E-01†(2.79E-02)**	1.24E-01(1.13E-02)	**2.16E-01†(2.34E-02)**
ER941	2.48E-02(4.94E-03)	**4.89E-02†(3.68E-03)**	9.53E-03(8.73E-04)	**1.68E-02†(1.29E-03)**
ER2344	5.41E-05§(7.84E-06)	**2.25E-04(6.17E-05)**	**6.47E-04(1.15E-05)**	6.01E-04≈(2.23E-05)
FF250	1.37E-01(5.51E-02)	**4.22E-01†(3.38E-02)**	2.68E-01(1.36E-02)	**5.62E-01†(4.81E-02)**
FF500	7.11E-02(2.79E-03)	**1.19E-01†(1.54E-02)**	3.88E-02(3.71E-03)	**6.09E-02†(7.46E-03)**
FF1000	4.96E-03(1.56E-04)	**1.72E-02†(3.06E-03)**	3.77E-03(7.26E-04)	**7.74E-03†(6.66E-04)**
FF2000	3.86E-04(3.45E-05)	**5.67E-04†(1.19E-05)**	4.22E-03(2.13E-04)	**9.89E-03†(6.41E-04)**

the metric values yielded by I-DMOEA-εC and I-MOEA/D. †, § and ≈ indicate that the performance of I-DMOEA-εC is better than, worse than, and similar to that of I-MOEA/D, according to the Wilcoxon's rank sum test, respectively. The bold data in each table are the best mean metric values for each instance.

As can be seen in Tables 6.4 and 6.5, in terms of *IGD* metric values, I-DMOEA-εC shows a significant advantage over I-MOEA/D on all test instances except for BA2500 with random weights, and WS500 and ER941 with logarithmic weights on which the two algorithms show competitive performance. As to the *HV*, I-DMOEA-εC shows significant superiority over I-MOEA/D on the majority of test instances. To be specific, I-MOEA/D performs better than I-DMOEA-εC on BA2500 with logarithmic weights, WS1000 with random and logarithmic weights, and ER2344 with random weights. The two algorithms perform competitvely on BA5000 and ER2344 with logarithmic weights. On the remaining test instances, I-DMOEA-εC outperforms I-MOEA/D significantly.

Figure 6.2 shows the distribution of the final solutions with the minimum *IGD* value within 20 runs found by I-MOEA/D and I-DMOEA-εC on BA2500, WS1500, ER2344 and FF2000 instances with random weights. It can be seen from Fig. 6.2 that I-MOEA/D and I-DMOEA-εC exhibit different behaviors and can find different parts of the PF for each instance. To be specific, I-MOEA/D tends to find solutions with high cost values, while I-DMOEA-εC shows an opposite behavior. Additionally, results obtained via I-DMOEA-εC achieve better convergence. The reason that I-DMOEA-εC cannot cover the PF well can be attributed to bad estimations of nadir points during the evolutionary process. In summary, Fig. 6.2 shows that I-MOEA/D and I-DMOEA-εC can find different parts of the PF for each instance and I-DMOEA-εC exhibits better performance compared with I-MOEA/D.

After optimizing the above-mentioned Bi-CNDP via certain MOEA, a set of non-dominated solutions will be obtained. We calculate the frequency of occurrence of each node in the set of final non-dominated solutions for all instances with random and logarithmic weights. Figure 6.3 shows the frequency values of all nodes that exist in the set of final non-dominated solutions for FF500 instances with random and logarithmic weights. According to Fig. 6.3, we note that frequency values of the majority of nodes of the instance with logarithmic weights are lower than that of the instance with random weights. This suggests that assigning higher weight values to nodes with larger node degrees has the effect of reducing frequency values of nodes in the non-dominated solutions. It can be concluded that it is an effective

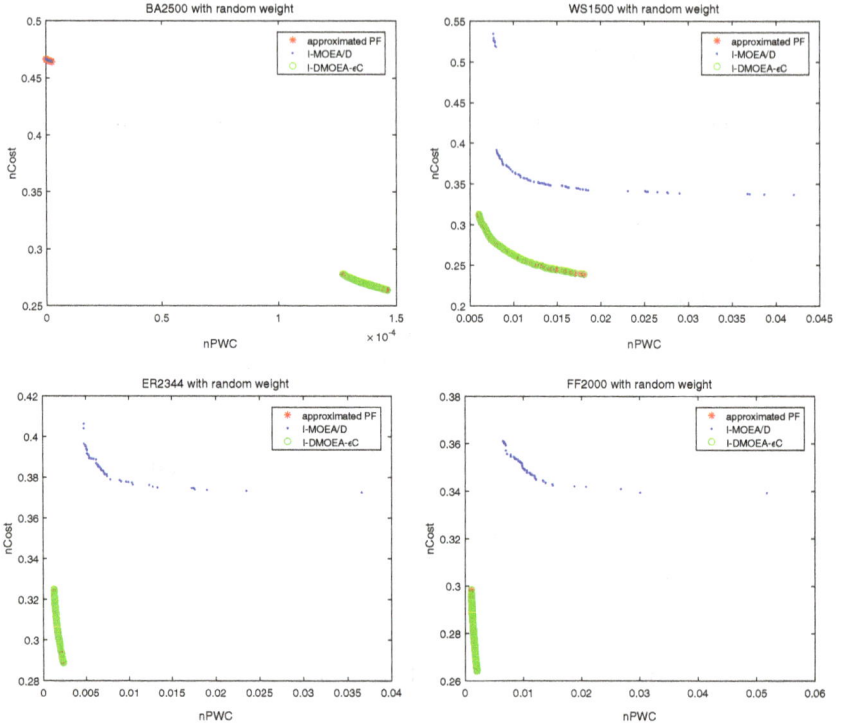

Fig. 6.2   Final populations in the objective space with the minimum *IGD* metric value within 20 runs obtained by I-MOEA/D and I-DMOEA-εC on BA2500, WS1500, ER2344 and FF2000 instances with random weights.

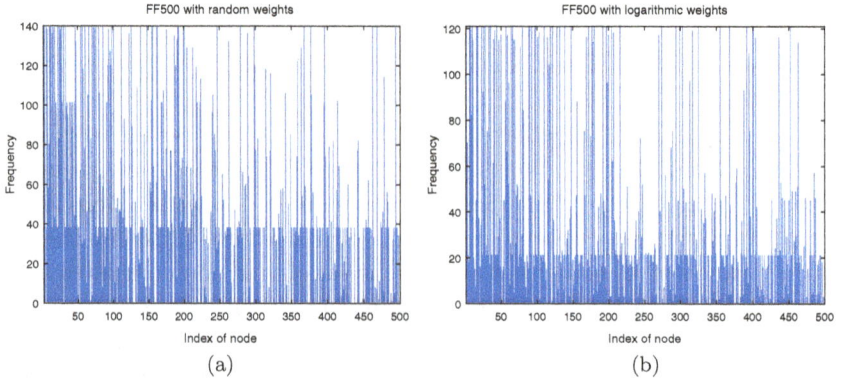

Fig. 6.3   Frequency of occurrence of each node in the set of final non-dominated solutions for FF500 instances with random and logarithmic weights.

way to spend more resources on protecting nodes with a high node degree in order to increase the robustness of a network.

Furthermore, a decision-making process from the perspective of minimizing the pairwise connectivity of the induced graph given a constraint on the cost of removing nodes is proposed. Specifically, after obtaining a set of non-dominated solutions, decision makers can select a preferred non-dominated solution with the smallest pairwise connectivity in the induced graph given a constraint on the cost of removing nodes. In the end, we exhibit the visual results of the decision-making process for the BA500, WS500, ER235 and FF500 instances, as shown in Figs. 6.4 to 6.7. In all

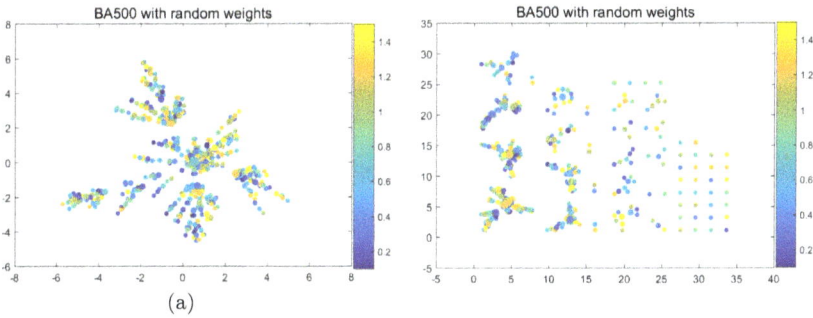

(a)

Fig. 6.4   The network of the BA500 instance with random weights: (a) initial; and (b) after removing a preferred non-dominated solution with the smallest pairwise connectivity in the induced graph and the total cost of removing nodes less than a predefined threshold 134. The size of each node is proportional to its degree, and the color of each node is related to its weight value.

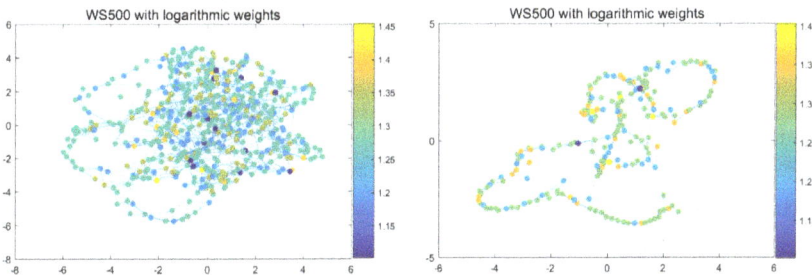

Fig. 6.5   The network of the WS500 instance with logarithmic weights: (a) initial; and (b) after removing a preferred non-dominated solution with the smallest pairwise connectivity in the induced graph and the total cost of removing nodes less than a predefined threshold 589. The size of each node is proportional to its degree, and the color of each node is related to its weight value.

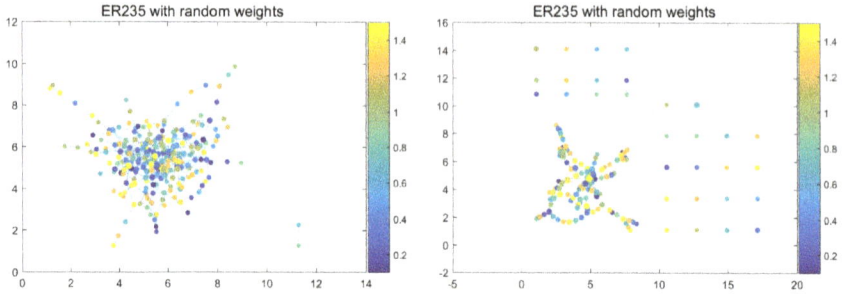

Fig. 6.6   The network of the ER235 instance with random weights: (a) initial; and (b) after removing a preferred non-dominated solution with the smallest pairwise connectivity in the induced graph and the total cost of removing nodes less than a predefined threshold 90. The size of each node is proportional to its degree, and the color of each node is related to its weight value.

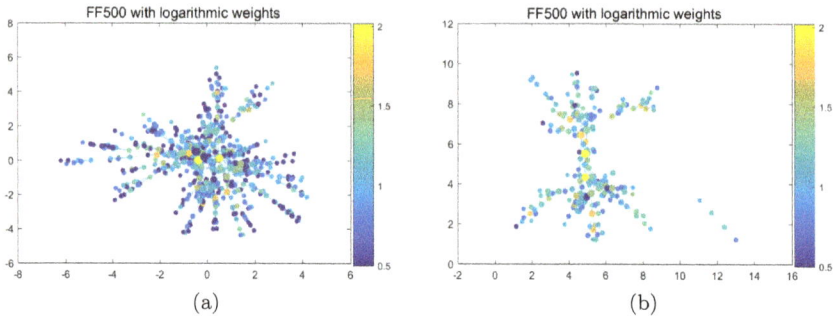

Fig. 6.7   The network of the FF500 instance with logarithmic weights: (a) initial; and (b) after removing a preferred non-dominated solution with the smallest pairwise connectivity in the induced graph and the total cost of removing nodes less than a predefined threshold 282. The size of each node is proportional to its degree, and the color of each node is related to its weight value.

figures, we use a colormap to represent the topology of the initial network and the network after removing a set of nodes based on a selected Pareto optimal solution. The size of each node is proportional to its degree, and the color of each node is related to its weight value.

6.5   Conclusion

Given a graph, the critical node detection problem consists of finding a set of nodes, the deletion of which satisfies one or more metrics in the induced graph. In contrast to most previous approaches, we use a bi-objective formulation, rather than make hypotheses on the psychology of decision makers. In this chapter, we propose and study a new variant of this problem called the bi-objective critical node detection problem (Bi-CNDP). In this formulation, we assume that removing each node has a cost, and decision makers want to minimize the pairwise connectivity of the induced graph and minimize the cost of removing a set of nodes at the same time.

We firstly prove the NP-hardness of this problem on general graphs and the existence of a polynomial algorithm for constructing the ε-approximated PF for Bi-CNDPs on trees. Then, different types of the mating pool and the replacement pool are proposed and integrated in two state-of-the-art decomposition-based MOEAs, including MOEA/D and DMOEA-εC. Two MOEAs and their variants are applied to solve the proposed Bi-CNDP. Sixteen common benchmark problems are modified by assigning random and logarithmic weight to each node. Computational experiments on all test instances were conducted first to evaluate the performance of different variants about the mating pool and the replacement pool. Then, further numerical experiments are used to compare the performance of two improved MOEAs, i.e., I-MOEA/D and I-DMOEA-εC, on the Bi-CNDP. Numerical results not only show the effectivenesses of the proposed fourth mating pool and the global replacement strategy, but also demonstrate different behaviors of two improved MOEAs and the superiority of I-DMOEA-εC on the majority of test problems. Finally, a decision-making process from the perspective of a single-objective is proposed for helping decision makers to identify the most critical nodes with the smallest pairwise connectivity and the total cost of removing nodes less than a predefined threshold.

Future research work will include investigations of designing more effective reproduction operators, embedding problem-specific knowledge during the optimization process, using single-objective methods to further refine solutions obtained via multi-objective approaches, and considering uncertainties in Bi-CNDPs.

Chapter 7

Solving Bi-objective Uncertain Stochastic Resource Allocation Problems by the CVaR-based Risk Measure and Decomposition-based Multi-objective Evolutionary Algorithms

This chapter investigates the uncertain stochastic resource allocation problem in which the results of a given allocation of resources are described as probabilities and these probabilities are considered to be uncertain from practical aspects. Here, uncertainties are introduced by assuming that these probabilities depend on random parameters that are impacted by various factors. The redundancy allocation problem (RAP) and the multi-stage weapon-target assignment (MWTA) problem are special cases of stochastic resource allocation problems. The bi-objective models for the uncertain RAP and MWTA problem by using the conditional value-at-risk (CVaR) measure to control the risk brought upon by uncertainties are built in this chapter. The bi-objective formulation covers the objectives of minimizing the risk of failure of completing activities and the resulting cost of resources. With the aim of determining referenced Pareto fronts (PFs), a linearized formulation and an approximated linear formulation are put forward for RAPs and MWTA problems based on problem-specific characteristics, respectively. Two state-of-the-art decomposition-based multi-objective evolutionary algorithms (i.e., MOEA/D-AWA and DMOEA-εC) are used to solve the formulated bi-objective problem. In view of the difference between MOEA/D-AWA and DMOEA-εC, two matching schemes inspired by DMOEA-εC are proposed and embedded in MOEA/D-AWA. Numerical experiments have been performed on a set of uncertain RAP and MWTA instances. Experimental results demonstrate that DMOEA-εC outperforms MOEA/D-AWA on the majority of test instances and the superiority of DMOEA-εC can be ascribed to the ε-constraint framework.

7.1   Introduction

The stochastic resource allocation problem is an extensive class of combinatorial optimization problems encountered in many fields, such as the radio resource allocation (RRA) problem in communication networks [Hao *et al.* (1998); Xu *et al.* (2014)], the relay resources management of wireless networks [Chu *et al.* (2016)], the circuit design [Li *et al.* (2016b)], the weapon target assignment (WTA) problem in military operations research fields [Blodgett *et al.* (2003); Li *et al.* (2016a)], the RAP in complex systems [Caserta and Voß (2016); Teimouri *et al.* (2016)], and so on. A distinct feature of the stochastic resource allocation problem is reflected by the fact that when assigning a number of resources to an activity, the completion of the activity is not definitely successful or failed, rather, it is stochastic and described as a probability value. For example, the reliability of a component is a stochastic event and is described as the reliability in the RAP and the completion of weapons-destroy-targets are stochastic events and is described as the kill probability in the WTA problem.

The risk of failure of completing activities and the resulting cost of resources are two commonly considered factors in stochastic resource allocation problems. In real situations, many conflicting criteria should be taken into account when evaluating the performance of decisions. Besides, many real-world applications are characterized by a host of uncertainties induced by incomplete, unobtainable and unquantifiable information. That is also the case for the stochastic resource allocation problem. It is then worth solving the stochastic resource allocation problem looking at both its multi-objective and uncertain aspects.

The RAP and the MWTA problems are special cases of stochastic resource allocation problems. They both have interested several researchers in the field of complex system. Previous approaches can be classified into two categories: exact and heuristic methods. The first category regroups methods that aim to end the optimal solution by ignoring the temporal constraints of the problem. The second category aims to provide sub-optimal solutions under temporal constraints. As reported in the literature review for this chapter, previous studies of the RAPs and the MWTA problems did not simultaneously consider the multiple objectives and uncertain aspects of the decision problem.

This chapter firstly investigates the uncertain stochastic resource allocation problem and presents generalized bi-objective formulations. Using terminologies of generalized models, models of two commonly encountered uncertain stochastic allocation problems, i.e., the RAP and the MWTA problem, are presented. Then the CVaR measure proposed in [Rockafellar and Uryasev (2000)] is adopted to measure the risk of failure of completing activities. Besides, two state-of-the-art decomposition-based multi-objective evolutionary algorithms (MOEAs) are employed to solve the formulated problems. The two decomposition-based MOEAs include the multi-objective evolutionary algorithms based on decomposition with adaptive weight adjustment (MOEA/D-AWA) [Qi *et al.* (2014)] and the decomposition-based multi-objective evolutionary algorithm with the ε-constraint framework (DMOEA-εC) [Chen *et al.* (2017)].

7.2   Literature Review

This section presents a brief literature review on the RAP, the MWTA problem and the uncertain optimization.

7.2.1   *The RAP*

System designers desire to achieve architectures and configurations with the highest possible reliability. In complex systems, to prevent severe consequences of system failure, improving products and systems reliability is vital. In general, there are two methods to increase system reliability, the first method is by increasing the reliability of components and the second one is by using redundant components within sub-systems [Tillman *et al.* (1980)]. The use of redundant components in sub-systems, which is called the system redundancy design, is the direct and the most common and efficient approach of enhancing the system reliability in industrial engineering activities. The RAP is one of the representative problems in system redundancy design. It seeks to determine the number of redundant components to be allocated in each sub-system with the purposes of improving the reliability of a system and reducing associated costs (e.g., acquisition, operation, maintenance costs, and so on) [Huang (2015)].

Since the RAP is known to be NP-hard [Chern (1992)], there are few exact algorithms that concentrate on techniques to reduce the search space of discrete optimization methods. Sung and Cho (2000) developed the lower and upper bounds of the system reliability by using variable relaxation and Lagrangean relaxation techniques. Prasad and Kuo (2000) presented a partial enumeration method for a wide range of complex optimization problems based on a lexicographic search.

Apart from exact methods, many heuristic algorithms have been proposed and applied in optimal reliability design. Nahas and Nourelfath (2005) combined a local search algorithm and a specific improvement algorithm that uses the remaining budget to improve the quality of a solution. Lee *et al.* (2002) developed a two-phase NN-hGA in which NN is used as a rough search technique to devise the initial solutions for a GA. Chen and You (2005) developed an immune algorithms-based approach inspired by the natural immune system of all animals. It analogizes antibodies and antigens as the solutions and objection functions, respectively. The RAP with a single objective has been extensively studied. Kaushik *et al.* (2013) presented a novel method to achieve the maximum reliability for the fault tolerant optimal network design problem. However, both factors are crucial for decision making in reality and should be taken into consideration. Zhu and Kuo (2014) dealt with a bi-objective RAP with the aims of maximizing the system reliability and minimizing the cost.

7.2.2   *The MWTA Problem*

The WTA problem is a fundamental problem arising in defense-related applications of military operations research [Ahuja *et al.* (2007)], which deals with how to obtain a weapon-target pair or a set of weapon-target pairs that meet decision makers' operational goals regarding combating effects and expenditures. The WTA problem is a classical constrained combinatorial optimization problem and has been proved to be NP-complete [Lloyd and Witsenhausen (1986)], which means any enumeration-based solver faces exponential computational complexity as the problem size increases. The research on WTA problems dates back to the 1950s and 1960s when Manne (1958), Braford (1961, AEREITM-9) and Day (1966) investigated the WTA modeling issues. Before the 1970s, research had been focused on the special areas, e.g., the missile-based aerial defense [Cai *et al.* (2006)]. Matlin summarized those works in [Matlin (1970)]. The general WTA problem was investigated systematically by Hosein and colleagues in the 1980s [Hosein *et al.* (1988)].

Hosein and Athans (1990a,b) grouped the WTA problems into two categories: the static WTA (SWTA) and the dynamic WTA (DWTA). In the SWTA, all weapons engage with targets in a single stage. On the contrary, the DWTA is a multi-stage problem where some weapons engage with targets at one stage, and the outcomes of this engagement are assessed. The strategy for the next stage is decided based on the former assessment. The DWTA is a global decision-making process that takes the whole defense effects through all stages into account and incorporates the concept of the time window, and thus it is much more complicated than the SWTA. In real combat situations, after making decisions at one stage, there will be a damage assessment process during which a number of new targets may appear or some old targets may exit [Wu *et al.* (2008)]. Following that, a new decision-making process of the next stage will be triggered. This process is the same as the previous one, except that the computational complexity is normally decreased due to the reduction of the number of weapons and targets. Thus, there is a cyclic computation: "Decision Making → Damage assessment → Decision Making" in the actual DWTA problem. The MWTA problem falls between the SWTA problem and the DWTA one. It also takes time windows into account, but does not possess the dynamic process. The MWTA problem, which lays a foundation for the DWTA, is considered in this chapter.

There are few cases in the literature of exact algorithmic solutions to the WTA due to its NP-completeness. The literature implementing exact solution techniques generally fall into one of two categories: small problems and problems wherein assumptions reduce the complexity [Kline *et al.* (2019)]. denBroeder *et al.* (1959) showed the first optimal solution technique in their Maximum Marginal Return (MMR) algorithm by assuming that the kill probability of each weapon to any target is the same. Several cases of using a branch and bound algorithm are found in the literature. Ahuja *et al.* (2007) implemented three lower bounding strategies for WTA problems, i.e., a generalized network flow solution, an MMR solution, and a minimum cost flow solution. Due to the computational complexity of the WTA problem, much of the literature focuses on heuristic algorithms that provide real time solutions rather than guaranteed optimal solutions. The very large scale neighborhood (VLSN) search metaheuristic is used to improve upon informed feasible solutions [Lee (2010)]. Their VLSN algorithms execute a heuristic search to efficiently find a quality solution and then they define local search neighborhoods within which to search for superior solutions.

Less attention has been given to the DWTA as compared to the SWTA. Thus, there are fewer heuristic algorithms shared among researchers. Often, hybrid heuristic algorithms are used to inform one another in execution. Xin *et al.* (2010) solve the DWTA problem using Virtual Permutation (VP), TS,

GA and ACO. In a subsequent work, they developed a rule-based heuristic which they consider the saturation of the constraints in order to inform the greedy selection process by which they assign weapons to targets in a stage t [Xin *et al.* (2011)]. A DWTA problem with the objectives of minimizing the operational cost and maximizing the expected damage to hostile targets was studied in [Li *et al.* (2015a)]. Leboucher *et al.* (2013) use the Hungarian Algorithm and a GA-PSO hybrid algorithm to determine the firing order of the assignments. A more recent review of WTA models and algorithms can be found in [Kline *et al.* (2019)].

7.3   Mathematical Formulations of the Uncertain Stochastic Resource Allocation Problem

In this section, general mathematical descriptions of the RAP and MWTA problem will be given.

7.3.1   *Mathematical Formulations of the RAP*

In general, redundancy strategies are generally grouped into two main categories, namely active and standby strategies. In an active redundancy strategy, it is assumed that all redundant components are implemented simultaneously. In a standby redundancy strategy, a sequential order is used for implementing the redundant components and these components can be added to the system at component failure time [Zhu and Kuo (2014)]. Besides, standby redundant systems can be broadly classified as having unrepairable or repairable elements. We focus on the unrepairable standby redundancy system here. A series-parallel system is basically characterized through a predefined number of sub-systems that are connected serially (see Fig. 7.1). A parallel sub-system works when at least one of its components works, and a series system fails when at least one of its sub-systems fails. The unrepairable series-parallel standby redundancy system is taken as an example in this chapter, and the following assumptions are used:

(1) Each component has only two possible states: functioning or failed;
(2) The states of all components are statistically independent; and
(3) All the components in a sub-system are identical.

Under the above-mentioned assumptions, the risk of failure of the series-parallel system with a confidence level α and the cost of the series-parallel

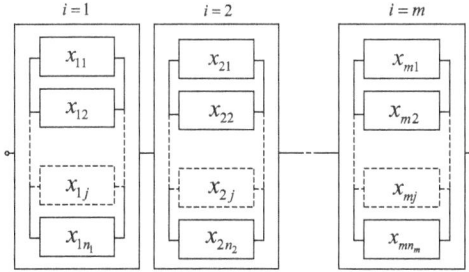

Fig. 7.1   The series-parallel system for RAPs.

system are defined as follows:

$$f_{sr}(X, \xi, \alpha) = CVaR_\alpha \left\{ \prod_{i=1}^{m} \prod_{j=1}^{n_i} (1 - r_{ij}(\xi))^{x_{ij}} \right\}, \qquad (7.1)$$

$$f_{sc}(X) = \sum_{i=1}^{m} \sum_{j=1}^{n_i} c_{ij} \cdot x_{ij}, \qquad (7.2)$$

where α represents the confidence level and m is the number of sub-systems. x_{ij} is the decision variable that represents the quantity of component j used in sub-system i. n_i is the number of available component choices for sub-system i. $r_{ij}(\xi)$ represents the uncertain reliability of component j in sub-system i and depends on a random parameter ξ. Different distributions of the reliability value $r_{ij}(\xi)$ can be available through expert knowledge before the system design. c_{ij} is the cost of component j in sub-system i.

Besides, there is a constraint on the maximum number of components for each sub-system. That is, for each sub-system i, the maximum number of components that can be parallelized is set as $n_{\max,i}$. This constraint can be formulated as following:

$$1 \leq \sum_{j=1}^{n_i} x_{ij} \leq n_{\max,i}, \quad i = 1, \ldots, m. \qquad (7.3)$$

Thus, the bi-objective uncertain RAP is modeled as: $\min[f_{sr}(X, \xi, \alpha), f_{sc}(X)]$, s.t. (7.3).

7.3.2   *Mathematical Formulations of the MWTA Problem*

Models of MWTA problems depend on many factors, e.g., defense strategies and features of targets and weapons. The scenario considered in this chapter is delineated as follows. At a certain time, the defender detects T hostile targets and has W weapons to intercept targets. Besides, before these offensive targets break through the defense and escape, there are at most S stages available for the defender to use its own weapons to hit hostile targets [Johnson *et al.* (2018); Li *et al.* (2015a, 2016a)]. The above-mentioned combat scenario is very common, e.g., in air-defense-oriented naval group combat. Given a set of targets and available weapons, one must find the optimal assignments of weapons to targets, considering both the risk of missing targets with a confidence level α and the ammunition cost of operations. The above-mentioned two factors are defined as follows:

$$f_{wr}(X^t, \xi, \alpha) = CVaR_\alpha \left\{ \sum_{j=1}^{T(t)} v_j \cdot \prod_{s=t}^{S} \prod_{i=1}^{W(t)} (1 - p_{ij}(s, \xi))^{x_{ij}(s)} \right\}, \quad (7.4)$$

$$f_{wc}(X^t) = \sum_{s=1}^{S} \sum_{j=1}^{T} \sum_{i=1}^{W} c_{ij}(s) \cdot x_{ij}(s), \quad (7.5)$$

where $f_{wr}(X^t, \xi, \alpha)$ in Eq. (7.4) is the CVaR value of missing targets at stage t with a confidence level α. f_{wc} in Eq. (7.5) is the overall ammunition consumption through all stages. $X^t = [X_t, X_{t+1}, \ldots, X_S]$ with $X_t = [x_{ij}(t)]_{W \times T}$ is the decision matrix at stage t, and $x_{ij}(t)$ is a binary decision variable taking a value of one (i.e., $x_{ij}(t) = 1$) if weapon i is assigned to target j at stage t, or zero (i.e., $x_{ij}(t) = 0$) otherwise. $W(t)$ and $T(t)$ represent the remaining number of the weapons and targets at stage t, respectively ($W(1) = W, T(1) = T$). v_j means that the threat value of target j. $p_{ij}(s, \xi)$ denotes the uncertain kill probability that weapon i destroys target j at stage s. It depends on a random parameter ξ that relates to the battle situation, weather conditions, and so on. Different distributions of the uncertain kill probability $p_{ij}(s, \xi)$ can be obtained, for example, by utilizing the historical observations of weapons' efficiency in different environments, or using simulated data, experts opinions, etc. Note that different probability distributions of $p_{ij}(s, \xi)$ represent different uncertain situations. $u_{ij}(s)$ denotes the ammunition consumption when weapon i is allocated to target j at stage s. β_i represents the unit economic cost of the ammunition that weapon i consumes. If all weapons are the same, β_i is assumed to be constant (e.g., one unit in this chapter).

In addition, there are some practical constraints on the usage of ammunitions as shown in the following:

$$\sum_{i=1}^{W} x_{ij}(t) \le m_j \quad \forall j \in I_j, \ \forall t \in I_t, \tag{7.6}$$

$$\sum_{j=1}^{T} x_{ij}(t) \le n_i \quad \forall i \in I_i, \ \forall t \in I_t, \tag{7.7}$$

$$\sum_{j=1}^{T} \sum_{t=1}^{S} x_{ij}(t) \le N_i \quad \forall i \in I_i, \tag{7.8}$$

$$x_{ij}(t) \le f_{ij}(t) \quad \forall i \in I_i, \ \forall j \in I_j, \ \forall t \in I_t, \tag{7.9}$$

$$I_i = \{1, 2, \dots, W\}; \quad I_j = \{1, 2, \dots, T\}; \quad I_t = \{1, 2, \dots, S\}.$$

Constraints (7.6) mean that at each stage, at most m_j ammunitions can be used to destroy target j, which limits the ammunition cost for each target at each stage. The value of m_j usually depends on the performance of available weapons and the tactical preferences of commanders. Constraints (7.7) reflect the capability of weapons firing at multiple targets at the same time. To be more precise, weapon i can fire at only most n_i targets at the same time. Actually, most weapons can fire at only one target, while for special cases, e.g., artillery-based defense systems, the value of n_i may be larger than two. In these cases, these weapons can be regarded as n_i independent weapons, so it is assumed that $n_i = 1$, $\forall i \in I_i$. Constraints (7.8) indicate the amount of available ammunitions of weapon i. Constraints (7.9) are engagement feasibility constraints that are important features of the MWTA against the SWTA, since the MWTA problem considers the influence of time windows on the engagement feasibility of weapons. If weapon i cannot shoot target j at stage s for various reasons (e.g., target j being beyond the range of weapon i), then $f_{ij}(t) = 0$; otherwise $f_{ij}(t) = 1$.

Based on the above, the formulation of the bi-objective uncertain MWTA problem is given as: $\min[f_{wr}(X^t, \xi, \alpha), f_{wc}(X^t)]$, s.t. (7.6)–(7.9).

7.4   Linearization of the RAP and the MWTA Problem

In this section, we linearize the RAP and approximate the MWTA problem with an integer programming problem. Thus, a true PF and a lower bound PF can be obtained for the RAP and MWTA problem, respectively.

7.4.1   *Linearization of the RAP*

Obviously, Eq. (7.1) can be easily converted into a linear function as described in the following:

$$f_{sr}^L(X, \xi, \alpha) = CVaR_\alpha \left\{ \sum_{i=1}^{m} \sum_{j=1}^{n_i} x_{ij} \cdot \ln(1 - r_{ij}(\xi)) \right\}. \qquad (7.10)$$

What's more, if a decision maker can express his/her preference by giving a threshold of the overall cost of a system (denoted as c_1), the second component of the bi-criteria objective function can be transformed into a constraint as following:

$$\sum_{i=1}^{m} \sum_{j=1}^{n_i} c_{ij} x_{ij} \le c_1. \qquad (7.11)$$

Based on above, the original bi-objective RAP can be transformed into a series of single objective problems by converting the objective regarding to the system cost into an additional constraint. What's more, the nonlinear objective regarding to the risk of failure of a system, i.e., $f_{sr}(X^t, \xi, \alpha)$, can be replaced by the linear function $f_{sr}^L(X^t, \xi, \alpha)$. We attempt to give a true PF of the original nonlinear bi-objective problem efficiently by solving a series of the following single objective linear problem: $\min f_{sr}^L(X^t, \xi, \alpha)$, s.t. (7.3) and (7.11).

7.4.2   *Linear Approximation of the MWTA Problem*

In Eq. (7.4), let $s_j = \prod_{s=1}^{S} \prod_{i=1}^{W} (1 - p_{ij}(s, \xi))^{x_{ij}(s)}$ and take logarithms on both sides, then we can obtain $\ln(s_j) = \sum_{s=1}^{S} \sum_{i=1}^{W} \ln(1 - p_{ij}(s, \xi)) \cdot x_{ij}(s)$. Let $y_j = -\ln(s_j)$ and observe the fact that $e^{-y_j} = \prod_{s=1}^{S} \prod_{i=1}^{W} (1 - p_{ij}(s, \xi))^{x_{ij}(s)}$. By introducing the term y_j in Eq. (7.4), we get the following formulation: $f_{wr}(X^t, \xi, \alpha) = CVaR_\alpha \{ \sum_{j=1}^{T} v_j \cdot e^{-y_j} \}$ [Ahuja *et al.* (2007)].

We consider $v_j \cdot e^{-y_j}$ at a series of values $\{y_j^1, y_j^2, \ldots, y_j^p\}$ and draw tangents of $v_j \cdot e^{-y_j}$ at these values. Let $F_j(p, y_j^k)$ $(k = 1, 2, \ldots, p)$ denote the upper envelope of these tangents. It is easy to see that the function $F_j(p, y_j^k)$ approximates $v_j \cdot e^{-y_j^k}$ from below, and $F_j(p, y_j^k)$ provides a lower bound on

$v_j \cdot e^{-y_j^k}$ for every y_j^k $(k = 1, 2, \ldots, p)$. Given the above, we can know that the function $F_j(p, y_j^k)$ gives a lower bound on the optimal objective function value for the MWTA problem. Specifically, the process of calculating the lower bound function $f_{wr}^L(X^t, \xi, \alpha)$ can be summarized as follows:

Step 1. First, we estimate the possible range of $y_j (j = 1, 2, \ldots, T)$, denoted as $[y_j^{min}, y_j^{max}]$, according to the data of test instances. Then, each range can be divided into $p - 1$ segments, i.e., $y_j^{min} = y_j^1 < y_j^2 < \cdots < y_j^{p-1} < y_j^p = y_j^{max}$, where p denotes the number of division points.

Step 2. The piecewise linear lower bound function can be obtained as following:

$$
f_{wr}^L(X^t, \xi, \alpha) = CVaR_\alpha \left\{ \sum_{j=1}^{T} v_j \cdot \max_{k=2,\ldots,p} \left\{ \theta_k \cdot \left(- \sum_{s=1}^{S} \sum_{i=1}^{W} \ln(1 - p_{ij}) \cdot x_{ij} \right. \right. \right.
$$
$$
\left. \left. \left. -y_j^{k-1} \right) + e^{-y_j^{k-1}} \right\} \right\}
$$
$$
y_j^{min} = y_j^1 < y_j^2 < \cdots < y_j^{p-1} < y_j^p = y_j^{max}, \ j = 1, \ldots, T
$$
$$
s.t. \quad \theta_k = \frac{e^{-y_j^k} - e^{-y_j^{k-1}}}{y_j^k - y_j^{k-1}}, \ k = 2, \ldots, p.
$$

(7.12)

Similar to the previous subsection, the overall cost of ammunitions can be transformed into a constraint by using a threshold of the cost (denoted as c_2), as shown in the following:

$$
\sum_{s=1}^{S} \sum_{j=1}^{T} \sum_{i=1}^{W} c_{ij}(s) x_{ij}(s) \leq c_2.
$$

(7.13)

Based on the above, the original bi-objective MWTA problem can be transformed into a series of single objective problems by converting the objective regarding to the ammunition cost into an additional constraint. What's more, the nonlinear objective regarding to the risk of missing targets, i.e., $f_{wr}(X^t, \xi, \alpha)$, can be bounded by the linear lower bound function $f_{wr}^L(X^t, \xi, \alpha)$. Thus we attempt to give a lower bound of the original nonlinear bi-objective problem efficiently by solving a series of following linear lower bound problems: min $f_{wr}^L(X^t, \xi, \alpha)$, s.t. (7.6)–(7.9) and (7.13).

7.5 Improved Decomposition-based MOEAs

Among various heuristics algorithms, the decomposition-based MOEAs have gained much attention. This chapter adopts two state-of-the-art decomposition-based multi-objective evolutionary algorithms (i.e.,

MOEA/D-AWA and DMOEA-εC) to deal with the RAP and the MWTA problem. This section presents a brief description of MOEA/D-AWA and DMOEA-εC at first and then proposes two matching schemes for MOEA/D-AWA in order to have a fair comparison with DMOEA-εC.

7.5.1 *Two Decomposition-based MOEAs*

Numerous methods including exact methods and intelligent optimization methods have been applied to solve resource allocation problems. Different exact approaches such as the Lagrangian relaxation and branch-and-bound methods have been developed in the literature to solve less complicated problems [Ahuja *et al.* (2007); Xin *et al.* (2011)]. Due to the high complexity induced by multiple objectives[1] and uncertainties,[2] exact methods cannot be employed to find optimal solutions for the model at hand in an acceptable time. As a result, two state-of-the-art decomposition-based MOEAs, including MOEA/D-AWA [Qi *et al.* (2014)] and DMOEA-εC [Chen *et al.* (2017)], are employed in this chapter. Additionally, MOEA/D-AWA is improved by adding two matching schemes at different stages of optimization.

Decomposition is an efficient and prevailing strategy for solving multiobjective optimization problems (MOPs). Its success was firstly witnessed by the multi-objective evolutionary algorithm based on decomposition (MOEA/D) [Zhang and Li (2007)] and its variants. MOEA/D decomposes an MOP into a set of scalar subproblems by using aggregated functions and uniformly distributed weight vectors and optimizes them concurrently. Commonly used aggregated functions used in MOEA/D include the weighted sum method, the Chebyshev method and the PBI method. Generally, the uniformity of weight vectors in MOEA/D can ensure the diversity of the optimal solutions based on the assumption that the PF is close to the hyper-plane in the objective space. However, the basic assumption might be violated in the case that the PF of the

---

[1]Since the single objective RAP and MWTA problems are both NP-hard [Chern (1992); Lloyd and Witsenhausen (1986)], the multi-objective formulation can only be harder to solve, which implies that the bi-objective RAP and MWTA problems are also NP-hard.
[2]For uncertain multi-objective optimization problems, multiple repeated function evaluations of a solution usually get different function values under different uncertain scenarios. Thus, it is difficult to definitely determine the quality of two solutions, which affects the ability of algorithms to find the optimum.

target MOP is complex. Therefore, some studies have been done to refine the weight vectors in MOEA/D [Jiang *et al.* (2011); Ma *et al.* (2014); Qi *et al.* (2014)]. MOEA/D-AWA [Qi *et al.* (2014)] is one of such research, in which an adaptive weight vector adjustment (AWA) strategy is introduced to obtain the uniformly distributed PF of the target MOP. It is natural to design an AWA strategy to regulate the distribution of weight vectors of subproblems periodically by removing subproblems from the crowded regions and adding new ones into the sparse regions, obtaining uniformly distributed non-dominated solutions consequently. Firstly, the AWA strategy removes subproblems located in the crowded regions whose crowded degrees are measured by the vicinity distance. Next, the elite population is deployed as a guidance to add new subproblems into the real sparse regions of the complex PF, rather than the discontinuous parts that are pseudo sparse regions. If an elite individual is located in a sparse region of the evolving population, it will be reused in the evolving population and a new weight vector will be generated and added to the subproblem set.

Algorithm 16 gives a brief description of MOEA/D-AWA. The notations used in MOEA/D-AWA are given in Table 7.1. Readers can refer to [Zhang and Li (2007)] and [Qi *et al.* (2014)] for a detailed description of MOEA/D and MOEA/D-AWA.

Table 7.1 Summary of Notations used in the Description of MOEA/D-AWA and DMOEA-εC.

m	The number of objective
N	The population size
δ	Probability of selecting mate solutions from its neighborhood
T	Neighborhood size
n_r	Maximum number of replacement
NFE	Maximum number of function evaluations
G_{max}	Maximum number of generations
$rate\_evol$	The ratio of iteration times to evolve with only MOEA/D
wag	The iteration interval of utilizing the adaptive weight vector adjustment strategy
num	Maximal number of subproblems needed to be adjusted
IN_m	The iteration interval of alternating the main objective function
$\mathbf{z}^*$	Ideal point
$\mathbf{z}^{nad}$	Nadir point
$rand$	A randomly distributed value in the interval $[0, 1]$

Algorithm 16 Framework of MOEA/D-AWA

1: **Input:** An MOP and related parameters.
2: **Output:** An external archive population EP.
3: Initialize N weight vectors by applying the WS-transformation on the original evenly spread weight vectors in MOEA/D; randomly initialize the evolving population $\mathbf{P} = \{\mathbf{x^1}, \ldots, \mathbf{x^N}\}$ and set $\mathbf{FV}^i = \mathbf{F}(\mathbf{x}^i)$; extract non-dominated individuals from P and denote the set of them as EP; initialize $\mathbf{z^*} = (z_1^*, \ldots, z_m^*)$ by setting $z_i^* = min\{f_i(\mathbf{x^1}), \ldots, f_i(\mathbf{x^N})\} - 10^{-7}$; set $gen = 0, n = N$.
4: **for** $i = 1$ to N **do**
5:    Set the neighborhood of the ith subproblem $B(i)$.
6:    $\pi^i = 1$.
7: **end for**
8: **while** $n \leq NFE$ **do**
9:    **for** $i \in I$ **do**
10:    $P = \begin{cases} B(i), & \text{if } rand < \delta \\ \{1, 2, \ldots, N\}, & \text{otherwise} \end{cases}$
11:    Reproduction: Select parent individuals from P randomly and apply certain reproduction operator to generate a new solution $\mathbf{y}$.
12:    $n = n + 1$.
13:    Repair: If $\mathbf{y}$ is infeasible, repair it.
14:    Update the approximated ideal point $\mathbf{z^*}$.
15:    Compare $\mathbf{y}$ with neighboring solutions of the subproblem i and update n_r neighboring solutions by using the aggregated objective value of each subproblem.
16:    Update the external archive EP and prune it by using the crowding distance-based approach.
17:    **end for**
18:    **if** $gen \leq rate\_evol * G_{max}$ and gen is a multiple of wag **then**
19:    Delete num overcrowded subproblems.
20:    Add num new subproblems into the sparse regions.
21:    **for** $i = 1$ to N **do**
22:       Find the T closest weights to ith subproblem and build new $B(i)$.
23:    **end for**
24:    **end if**
25:    $gen = gen + 1$.
26: **end while**

Apart from the above-mentioned aggregation methods, the ε-constraint method is also a basic generation method in mathematical programming [Miettinen (1999)] and is often used as an element of more developed methods. DMOEA-εC takes inspiration from the ε-constraint method and is a newly proposed multi-objective solver in recent research [Chen *et al.* (2017)]. The ε-constraint method selects one of the objectives as the main objective, while transforming the other non-main objectives to constraints and associating each non-main objective with an upper bound coefficient. DMOEA-εC explicitly decomposes an MOP into a series of scalar-constrained optimization subproblems by selecting one of the objectives as the main objective function and associating each subproblem with an upper bound vector. These subproblems are optimized collaboratively by an evolutionary algorithm based on the feasibility rule [Deb (2000)]. Besides, each subproblem is optimized using information only from its neighboring subproblems.

Under the ε-constraint framework, DMOEA-εC tends to retain feasible solutions for each subproblem, which will be bad for the optimization of the main objective function. A main objective alternation strategy is proposed to tackle this problem periodically. After the main objective alternation strategy is utilized, a solution which is good for the current subproblem may not perform well since the objective function of this subproblem has been changed. Thus a solution-to-subproblem matching scheme is designed to place the nearest solution to each subproblem. It uses the distance value between a solution and a subproblem as the criterion and is utilized after the main objective alternation strategy. Additionally, when a new solution is generated, it may perform badly for the current subproblem but perform well for another subproblem. In order to avoid wasting potentially useful solutions and make the best use of them, the subproblem-to-solution matching scheme, which uses the constraint violation value as the criterion is proposed to find a subproblem with the minimum constraint violation value for the new solution. The two matching schemes strike a balance between convergence and diversity. DMOEA-εC has been compared with six state-of-the-art MOEAs on both continuous and 0/1 knapsack benchmark problems and has shown its advantages over competitive approaches. Given parameters presented in Table 7.1, the DMOEA-εC is summarized in **Algorithm 1**. Readers can refer to [Chen *et al.* (2017)] for a detailed description of DMOEA-εC.

As stated in [Chen *et al.* (2017)], the superiority of DMOEA-εC can be partly attributed to the effectivenesses of two matching schemes, which are capable of balancing convergence and diversity at different stages of evolution. Numerical results on analyzing algorithmic behaviors of DMOEA-εC have confirmed this statement. Then are the two matching schemes applicable to MOEA/D-AWA, which is also decomposition-based? If so, can two matching schemes enhance the performance of MOEA/D-AWA? In order to answer these questions, we design two matching schemes for MOEA/D-AWA and examine their performance.

7.5.2   *Two Matching Schemes*

Although   MOEA/D-AWA   and   DMOEA-εC   are   different algorithms, they share the similar concept of decomposition. In MOEA/D-AWA, there are two types of distance, i.e., d_1 and d_2. Suppose L is a line passing through the ideal point $\mathbf{z}^*$ with a direction $\mathbf{w}$, $d_1 = \frac{\left\| (\mathbf{z}^* - \mathbf{F}(\mathbf{x}))^T \mathbf{w} \right\|}{\|\mathbf{w}\|}$ is the distance between $\mathbf{z}^*$ and the projection of $\mathbf{F}(\mathbf{x})$ on the line L. $d_2 = \left\| \mathbf{F}(\mathbf{x}) - (\mathbf{z}^* + d_1 \frac{\mathbf{w}}{\|\mathbf{w}\|}) \right\|$ is the perpendicular distance from $\mathbf{F}(\mathbf{x})$ to the line L. d_1 is used to evaluate the convergence of $\mathbf{x}$ toward the PF and d_2 is a kind of measure for the diversity of a population.

Inspired by DMOEA-εC, d_2 can firstly be adopted as the matching criterion for the solution-to-subproblem matching scheme to find the solution with minimum perpendicular distance value for each subproblem. The solution-to-subproblem matching scheme is used periodically and it can place the nearest solution to each subproblem at a large degree, which is beneficial to diversity. The subproblem-to-solution matching scheme utilizes d_1 to match the subproblem with the smallest d_1 value for a newly generated solution in order to make the following replacement more effective. Additionally, the subproblem-to-solution matching scheme is adopted after a new solution is generated, which is good for convergence. Two matching schemes are expected to force solutions to stay close to certain weight vectors and explicitly maintain the desired distribution of solutions in the evolutionary process, which leads to a balance between convergence and diversity in multi-objective optimization. MOEA/D-AWA with the above-mentioned two matching schemes is denoted as MOEA/D-AWA-M and will be compared with MOEA/D-AWA to verify the effectiveness of two matching schemes.

In addition, the genetic algorithm is adopted as the search engine in MOEA/D-AWA, MOEA/D-AWA-M and DMOEA-εC. In this case, two-point crossover, one-point mutation and random repair mechanism are used for both bi-objective RAP and MWTA problems.

7.6 Computational Experiments

This section is devoted to the experimental design for investigating the effectiveness of two matching schemes designed for MOEA/D-AWA and the performance of two multi-objective solvers on solving uncertain RAP and MWTA instances. First, descriptions of the test instances of the RAP and MWTA are given. Then parameter tunings for both the RAP and MWTA instances are conducted. Finally, experimental results are illustrated.

7.6.1 *Test Instances*

As to RAPs, two well-known benchmark cases that consist of 3 sub-systems with 5 components and 14 sub-systems with 4 components are considered [Khalili-Damghani and Amiri (2012); Li *et al.* (2009); Taboada *et al.* (2007)]. The two benchmark cases are denoted as RAP1 and RAP2, respectively. The maximum number of components that can be paralleled for each sub-system $n_{\max}$ is set as $n_{\max} = [7\,7\,7]$ and $n_{\max} = [6\,5\,6\,6\,6\,6\,5\,6\,6\,6\,5\,6\,6\,5]$ for RAP1 and RAP2, respectively. Tables 7.2 and 7.3 present the information concerning RAP1 and RAP2.

For MWTA problems, since the actual data is difficult to obtain, three different-scaled instances, i.e., small, medium and large-scaled instances, are randomly generated. The numbers of weapons/targets/stages in small, medium and large-scaled instances are 3/5/3, 20/12/5 and 50/50/8, respectively [Li *et al.* (2015a, 2016a)]. The value vector of targets comes from [Cai (2010)]. For each instance, each element of the deterministic kill probability and ammunition consumption matrices at stage s, namely $P(s) = [p_{ij}(s)]_{W \times T}$ and $U(s) = [u_{ij}(s)]_{W \times T}$, is randomly generated within a given range related to the problem scale. Parameters in constraints including m_i, n_i and N_i are deterministic.

Table 7.2   Data for the RAP Benchmark Case 1 (RAP1).

Sub-system i	Component Type j									
	1		2		3		4		5	
	R	C	R	C	R	C	R	C	R	C
1	0.94	9	0.91	6	0.89	6	0.75	3	0.72	2
2	0.97	12	0.96	3	0.70	2	0.66	2	—	—
3	0.96	10	0.86	6	0.72	4	0.71	4	0.67	2

Table 7.3   Data for the RAP Benchmark Case 2 (RAP2).

Sub-system i	Component Type j							
	1		2		3		4	
	R	C	R	C	R	C	R	C
1	0.90	1	0.93	1	0.91	2	0.95	2
2	0.95	2	0.94	1	0.93	1	—	—
3	0.85	2	0.90	3	0.87	1	0.92	4
4	0.83	3	0.87	4	0.85	5	—	—
5	0.94	2	0.93	2	0.95	3	—	—
6	0.99	2	0.98	3	0.97	2	0.96	2
7	0.91	4	0.92	4	0.94	5	—	—
8	0.81	3	0.90	5	0.91	6	—	—
9	0.97	2	0.99	3	0.96	6	—	—
10	0.83	4	0.85	4	0.90	5	—	—
11	0.94	3	0.95	4	0.96	5	—	—
12	0.79	2	0.82	3	0.85	4	0.90	5
13	0.98	2	0.99	3	0.97	2	—	—
14	0.90	4	0.92	4	0.95	5	0.99	6

As to uncertain RAPs and MWTA problems, in order to generate uncertain instances based on the deterministic instances, a common probability distribution that the uncertain completion probability may follow can be applied over the deterministic data. Specifically, each uncertain completion probability, namely $prob_{ij}(\xi)$,[3] is sampled from the following uniform distribution:

$$prob_{ij}(\xi) \sim U((1 - \gamma) \cdot prob_{ij}, (1 + \gamma) \cdot prob_{ij}), \qquad (7.14)$$

where the parameter γ is used to tune the degree of the deviation of the completion probability. In any case, the central tendency of the distribution always corresponds to a deterministic completion probability value $prob_{ij}$. In the following, γ is set as 0.2. As to CVaR, α stands for the confidence level and is set as $\alpha = 0.9$.

7.6.2   *Encoding*

An appropriate solution encoding scheme is crucial for an algorithm when solving optimization problems. For both RAPs and MWTA problems, the

---

[3] $prob_{ij}(\xi)$ equals to $r_{ij}(\xi)$ and $p_{ij}(s, \xi)$ for RAPs and MWTA problems, respectively.

(a)

(b)

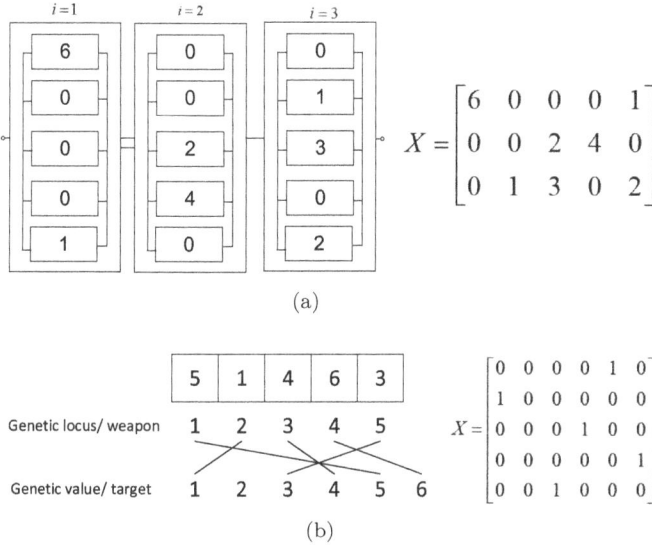

Fig. 7.2   Examples of the encoding scheme of (a) the RAP problem (3 sub-systems with 5 components), and (b) the MWTA problem ($W = 5, T = 6, S = 1$).

integer encoding is adopted. Every solution of the RAP is represented by a matrix, and each element of the matrix is the number of components used in each sub-system. As to the MWTA problem, each solution is represented by a vector whose length is the number of weapons. Each weapon is regarded as a genetic locus to form a chromosome, and the genetic value of each genetic locus indicates the number of the target to which the weapon is assigned. Such an encoding method can guarantee that every solution satisfies the constraint (7.7) naturally. As to the other constraints, it will be checked whether each new solution satisfies these constraints or not. If not, this solution should be repaired according to a random repair mechanism. Figure 7.2 illustrates examples of the encoding scheme of the RAP problem (3 sub-systems with 5 components) and the MWTA problem ($W = 5, T = 6, S = 1$).

7.6.3   *Performance Metrics*

Two commonly used performance metrics, i.e., inverted generational distance (IGD) [Zhou *et al.* (2005)] and hypervolume (HV) [Zitzler and Thiele (1999)] are employed to evaluate the performance of all compared algorithms.

7.6.4   *Parameter Tuning*

Since different levels of the parameters have effects on the quality of solutions obtained by certain algorithms, the Taguchi procedure [Montgomery (2005); Roy (2010)] has been frequently used to tune parameters of algorithms. The Taguchi procedure is applied in the field of design-of-experiments (DOE) and is regarded as an efficient alternative for full factorial experiments. This procedure uses a special orthogonal table for setting a set of experiments to investigate a group of parameters. The preferred parameter settings are then determined through analysis of the signal-to-noise (SN) ratio where factor levels that maximize the appropriate SN ratio are optimal. The SN ratio is defined as:

$$SN = -10 \times \log\left(\frac{1}{n}\sum_{i=1}^{n} y_i\right), \tag{7.15}$$

where y_i and n are the response value and the number of orthogonal arrays in the orthogonal table, respectively. In MOEAs, two main goals (convergence and diversity) are considered simultaneously. As mentioned above, both IGD and HV are suitable metrics. However, the Taguchi method only deals with one response function. Therefore, a combination of the performance measures should be defined. In this study, an integrated metric, defined as IGD/HV, will act as the response variable of the Taguchi method.

With the purpose of calculating the IGD metric value, the true Pareto front of RAP instances and the lower bound PF of MWTA instances generated through the linearized formation described in Section 7.4 will be chosen as P^*. The reference points for each instance used in calculations of the HV metric values are set as 1.1 times the true nadir point based on P^*.

Since the processes of the Taguchi procedure are similar across different approaches, we only display the detailed analyses process of MOEA/D-AWA. To be specific, the first stage of the Taguchi method is to determine the factors and their levels. The common parameters that MOEA/D-AWA, MOEA/D-AWA-M and DMOEA-εC share are the population size (N), the number of iterations (Num), the crossover rate (CR), and the mutation rate (MR). For a fair comparison, the number of function evaluations $(NFE = N \cdot Num)$ is set the same for each instance. Thus, the factors to be tuned include: the combination of the population size and the number of iterations (N, Num), the crossover rate CR, and the mutation rate MR. Different levels of these factors are shown in Table 7.4.

Table 7.4   Algorithm Parameters and their Levels for RAP and MWTA Instances.

Parameters	Factor Level		
	1	2	3
RAP1 (N, Num)	(50,400)	(100,200)	(200,100)
RAP2 (N, Num)	(50,600)	(100,300)	(300,100)
Small-scaled (N, Num)	(50,400)	(100,200)	(200,100)
Medium-scaled (N, Num)	(80,500)	(100,400)	(200,200)
Large-scaled (N, Num)	(200,500)	(100,1000)	(500,200)
CR	0.5	0.7	0.9
MR	0.01	0.05	0.1

Since three factors with three levels exist, the proper orthogonal table is $L_9(3^3)$. MOEA/D-AWA is applied to all combinations of factors, and response values are calculated. The orthogonal table of these designs along with the response values for the RAP and the MWTA instances of MOEA/D-AWA are shown in Table 7.5. Readers can refer to [Montgomery (2005); Roy (2010)] for more details on the Taguchi approach.

According to the orthogonal table, main effect plots of the SN ratio for RAP and MWTA instances of MOEA/D-AWA are illustrated in Figs. 7.3 and 7.4. Then, we figure out the change of response values and analyze the significance of the rank of each parameter and obtain the response table as shown in Table 7.6. It can be seen that: (N, Num) is the parameter that has the most significant impact on the algorithm for most RAP and MWTA instances. According to the response table, the best combinations of the parameter values for each instance are determined. Similar processes can be applied to MOEA/D-AWA-M and DMOEA-εC. The best combination of parameters for the RAP and MWTA instances of all three approaches are listed in Table 7.7.

In addition to these common parameters, the remaining parameters, including the neighborhood size T, the maximal number of replacement n_r, the maximal number of adjusted subproblems num, and the iteration interval of utilizing the dynamic resource allocation strategy and performing the solution-to-subproblem matching procedure IN_m, are also shown in Table 7.7. Besides, the probability of selecting mate solutions from the neighborhood is set as $\delta = 0.9$, the ratio of the iterations to evolve with only MOEA/D is set as $rate\_evol = 0.7$ [Qi et al. (2014)], and the size of the external population is set as $2N$ for all test instances.

Table 7.5 Orthogonal Table for RAP and MWTA Instances of MOEA/D-AWA.

Experiment Number	Factors			Response Value				
	(N, Num)	CR	MR	RAP1	RAP2	Small-scaled	Medium-scaled	Large-scaled
1	1	1	1	9.1514	3.6937	2.0622	6.6456E−03	1.0213E+02
2	1	2	2	9.4234	4.7038	3.0797	6.1089E−03	1.0848E+02
3	1	3	3	9.6815	3.1249	2.5660	6.3931E−03	1.3519E+02
4	2	1	2	9.2761	3.5207	2.2045	6.1269E−03	9.9797E+01
5	2	2	3	8.4553	3.4335	2.2145	6.1620E−03	9.5280E+01
6	2	3	1	7.9091	3.9401	1.7004	6.1975E−03	6.2556E+01
7	3	1	3	9.6724	2.9148	2.2979	6.7583E−03	1.0998E+02
8	3	2	1	9.1355	3.8365	1.9409	6.3235E−03	1.1252E+02
9	3	3	2	9.5272	4.5699	2.4670	6.4101E−03	1.3952E+02

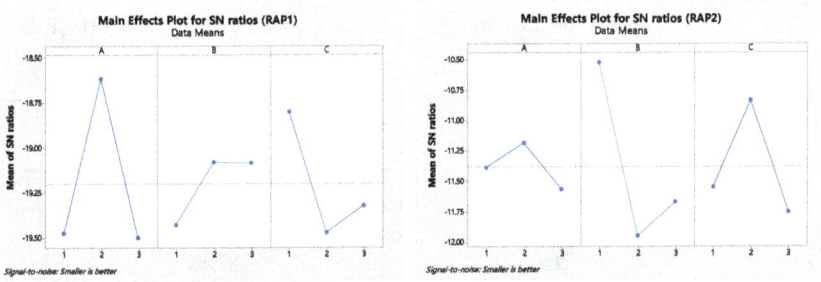

Fig. 7.3   Main effect plots for RAP instances of MOEA/D-AWA.

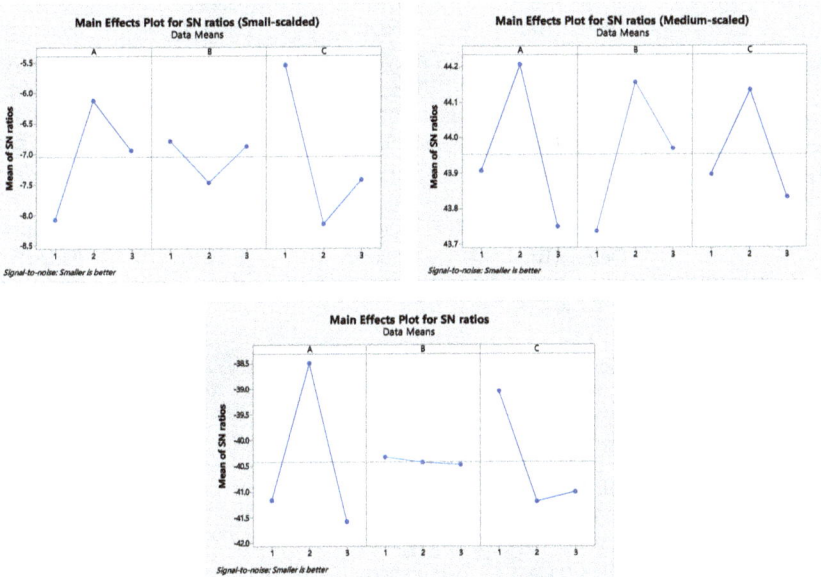

Fig. 7.4   Main effect plots for MWTA instances of MOEA/D-AWA.

7.6.5   *Experimental Results*

This section performs experiments to examine the effects of the two matching schemes proposed for MOEA/D-AWA and compares the performance
of two multi-objective approaches on solving RAP and MWTA instances.
For this part of experiments, the above-mentioned performance metrics,
i.e., *IGD* and *HV*, are employed to evaluate the performance of compared
algorithms.

Table 7.6   Response Table for the RAP and MWTA Instances of MOEA/D-AWA.

	Level	(N, Num)	CR	MR
RAP1	1	9.419	9.367	8.732
	2	8.547	9.005	9.409
	3	9.445	9.039	9.270
	Delta. (Rank)	0.898(1)	0.362(3)	0.677(2)
RAP2	1	3.204	3.245	3.311
	2	3.166	3.243	3.227
	3	3.328	3.210	3.161
	Delta. (Rank)	0.162(1)	0.035(3)	0.150(2)
Small-scaled	1	2.569	2.188	1.901
	2	2.040	2.412	2.584
	3	2.235	2.244	2.359
	Delta. (Rank)	0.529(2)	0.224(3)	0.683(1)
Medium-scaled	1	6.383E-03	6.510E-03	6.389E-03
	2	6.162E-03	6.198E-03	6.215E-03
	3	6.497E-03	6.334E-03	6.438E-03
	Delta. (Rank)	0.000335(1)	0.000312(2)	0.000223(3)
Large-scaled	1	115.27	103.97	92.40
	2	85.88	105.43	115.93
	3	120.67	112.42	113.48
	Delta. (Rank)	34.80(1)	8.45(3)	23.53(2)

Table 7.7   The Best Combination of Parameters for the RAP and the MWTA Instances of 3 Algorithms.

Instances	Approaches	(N, Num)	CR	MR	T	n_r	num	IN_m
RAP1	MOEA/D-AWA	(100,200)	0.7	0.01	5	1	1	—
	MOEA/D-AWA-M	(100,200)	0.7	0.1	5	1	1	20
	DMOEA-εC	(100,200)	0.7	0.1	5	1	—	20
RAP2	MOEA/D-AWA	(100,300)	0.5	0.05	8	3	3	—
	MOEA/D-AWA-M	(100,300)	0.7	0.1	8	3	3	30
	DMOEA-εC	(100,300)	0.9	0.1	8	3	—	30
Small-scaled	MOEA/D-AWA	(100,200)	0.5	0.01	3	1	1	—
	MOEA/D-AWA-M	(100,200)	0.5	0.01	3	1	1	40
	DMOEA-εC	(100,200)	0.5	0.01	3	1	—	40
Medium-scaled	MOEA/D-AWA	(100,400)	0.7	0.05	10	3	3	—
	MOEA/D-AWA-M	(100,400)	0.5	0.1	10	3	3	10
	DMOEA-εC	(100,400)	0.7	0.05	10	3	—	10
Large-scaled	MOEA/D-AWA	(100,1000)	0.5	0.01	15	5	5	—
	MOEA/D-AWA-M	(100,1000)	0.5	0.01	15	5	5	20
	DMOEA-εC	(100,1000)	0.9	0.01	15	5	—	20

As mentioned above, since experimental results have confirmed that the performance of DMOEA-εC is enhanced because of the introduction of two matching schemes, two similar matching schemes are proposed and embedded in MOEA/D-AWA. Therefore, effects of the two proposed matching schemes should be examined on all test instances by comparing the performance of MOEA/D-AWA with MOEA/D-AWA-M.

Besides, numerical experiments are also conducted to demonstrate the performance of DMOEA-εC on solving RAP and MWTA instances.

The means and standard deviations of *IGD* and *HV* metric values over 30 runs of MOEA/D-AWA, MOEA/D-AWA-M and DMOEA-εC on RAP and MWTA instances are shown in Tables 7.8 and 7.9, respectively. The numbers in parentheses are the standard deviations. Besides, the mean *HV* (*IGD*) values for each instance are sorted in descending (ascending) order, and the numbers in the square brackets are their ranks. The Wilcoxon's rank sum test at a 5% significance level is conducted to test the significance of differences between DMOEA-εC and the other algorithms. †, § and ≈ indicate the performance of the comparison algorithm is better than, worse than, and similar to that of DMOEA-εC according to the Wilcoxon's rank sum test, respectively. The bold data in each table are the best mean metric values for each instance. The last row of each table presents the mean rank value of each algorithm over all test instances.

As can be seen in Table 7.8, in terms of *IGD* metric values, MOEA/D-AWA-M shows obvious advantages over MOEA/D-AWA on the majority of test instances. Similar results can be obtained in terms of *HV* values as shown in Table 7.9. This observation demonstrates that two proposed matching schemes are helpful for improving the performance of MOEA/D-AWA on solving RAP and MWTA instances. Effects of the two matching schemes in MOEA/D-AWA are similar to that in DMOEA-εC. That is, the solution-to-subproblem matching scheme using d_2 as a matching criterion can place the nearest solution to each subproblem at a large degree, which is beneficial to keep a diverse population. The subproblem-to-solution matching scheme with d_1 as the matching criterion is able to avoid wasting potentially useful solutions and make best use of new solutions, and thus it is good for convergence. Therefore, two matching schemes strike a balance between diversity and convergence.

As can be seen in Tables 7.8 and 7.9, in terms of *IGD* metric values, DMOEA-εC shows significant advantage over MOEA/D-AWA-M on all test instances except for RAP1 and small-scaled MWTA instances on which two algorithms show competitive performance. As to

Table 7.8   Mean *IGD* Metric Values of MOEA/D-AWA and DMOEA-εC and their Variants about the 2 Matching Schemes and the Hierarchical Comparison Strategy over 30 Independent Runs on RAP and MWTA Instances with the Noise Strength $\gamma = 0.2$ and the Confidence Value $\alpha = 0.9$.

	MOEA/D-AWA	MOEA/D-AWA-M	DMOEA-εC
RAP1	8.8691E-00(1.2312E-02)§[3]	8.7325E-00(1.1307E-02)≈[2]	**8.6655E-00**(3.3665E-02)[1]
RAP2	2.6850E-00(8.8345E-03)§[3]	2.5753E-00(1.9516E-02)§[2]	**2.4990E-00**(2.0992E-02)[1]
Small-scaled	1.4510E-00(4.9607E-01)§[3]	**1.3013E-00**(4.8308E-01)≈[1]	1.3076E-00(4.0791E-01)[2]
Medium-scaled	1.4245E+01(6.8510E-02)§[2]	1.3230E+01(2.5467E-03)§[3]	**2.8587E-00**(6.3086E-02)[1]
Large-scaled	3.7567E+01(1.6531E-01)§[3]	3.1206E+01(1.9144E-01)§[2]	**2.4253E+01**(1.0463E-01)[1]
Mean rank	2.75	2.00	1.25

Table 7.9   Mean *HV* Metric Values of MOEA/D-AWA and DMOEA-εC and their Variants about the 2 Matching Schemes and the Hierarchical Comparison Strategy over 30 Independent Runs on RAP and MWTA Instances with the Noise Strength $\gamma = 0.2$ and the Confidence Value $\alpha = 0.9$.

	MOEA/D-AWA	MOEA/D-AWA-M	DMOEA-εC
RAP1	9.2501E-01(2.3135E-02)§[3]	9.2311E-01(6.5411E-02)≈[2]	**9.3812E-01**(2.5521E-02)[1]
RAP2	7.1106E-01(3.4008E-02)§[3]	7.7913E-01(3.2662E-02)§[2]	**7.8221E-01**(4.9568E-02)[1]
Small-scaled	4.3556E-01(3.0310E-03)§[3]	4.5783E-01(1.2571E-03)§[2]	**4.7540E-01**(7.8115E-04)[1]
Medium-scaled	2.1436E+03(2.3110E-01)§[3]	2.2579E+03(6.7151E-02)§[2]	**2.5671E+03**(2.5275E-01)[1]
Large-scaled	3.3680E-01(1.9591E-02)§[3]	3.5078E-01(2.3849E-02)≈[2]	**3.8771E-01**(1.4577E-02)[1]
Mean rank	3.00	2.00	1.00

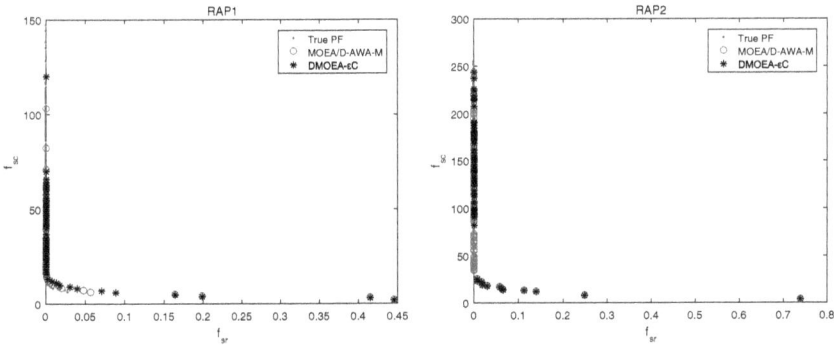

Fig. 7.5 Final populations in the objective space with the minimum *IGD* metric value within 30 runs obtained by MOEA/D-AWA-M and DMOEA-εC on RAP test instances.

the *HV*, DMOEA-εC shows significant superiority over MOEA/D-AWA-M on the majority of test instances. To be specific, DMOEA-εC performs better than MOEA/D-AWA-M on RAP2, small-scaled MWTA, and medium-scaled MWTA instances. Two algorithms perform similarly on the RAP1 instance. Tables 7.8 and 7.9 also present the overall rank of the three algorithms, that is DMOEA-εC, MOEA/D-AWA-M and MOEA/D-AWA, according to the mean rank values.

Figures 7.5 and 7.6 show the distribution of the final population in the objective space with the minimum *IGD* value within 30 runs found by MOEA/D-AWA-M and DMOEA-εC on RAP and MWTA instances, respectively. It can be seen from Figs. 7.5 and 7.6 that MOEA/D-AWA-M and DMOEA-εC exhibit similar performance on RAP2 and the small-scaled MWTA instance. For the RAP1 instance, MOEA/D-AWA-M and DMOEA-εC omit a small part of the true PF. As to the medium-scaled MWTA instance, DMOEA-εC achieves better convergence than MOEA/D-AWA-M. In this case, both MOEA/D-AWA-M and DMOEA-εC cannot cover the approximated PF well. This phenomenon can be attributed to bad estimations of the ideal point and the nadir point during the evolutionary process. In summary, Figs. 7.5 and 7.6 display that MOEA/D-AWA-M and DMOEA-εC can approximate the PF very well for the majority of instances and DMOEA-εC exhibits better performance as compared with MOEA/D-AWA-M.

In conclusion, the experimental results on MWTA and RAP instances indicate that the superiority of DMOEA-εC is significant on RAP2 and the medium-scaled MWTA test instances, but not significant on RAP1 and

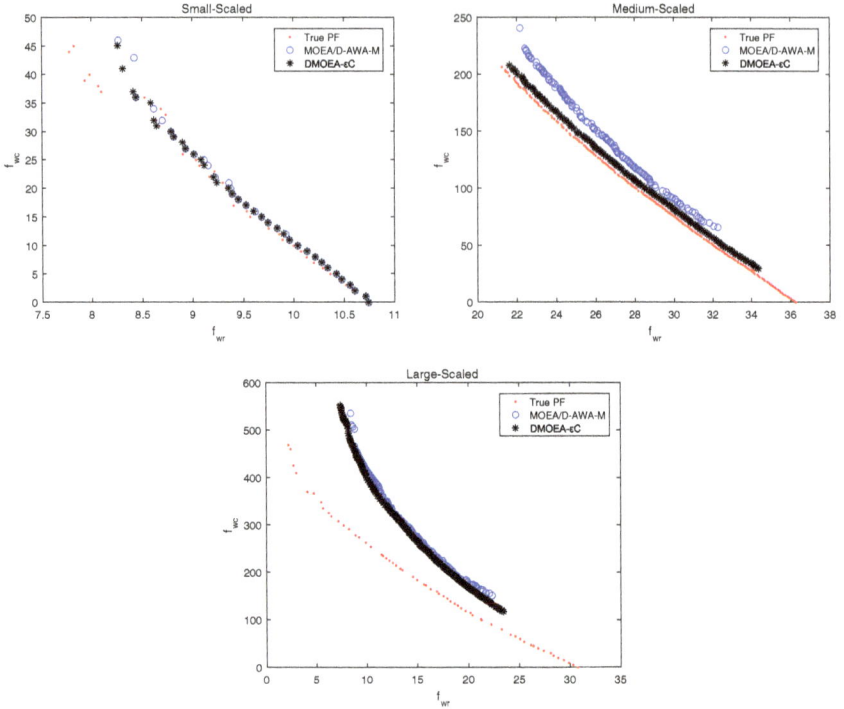

Fig. 7.6 Final populations in the objective space with the minimum *IGD* metric value within 30 runs obtained by MOEA/D-AWA-M and DMOEA-εC on MWTA test instances.

small-scaled MWTA test instances. The main difference between MOEA/D-AWA-M and DMOEA-εC lies in their algorithmic framework, and the superiority of DMOEA-εC against MOEA/D-AWA-M can be ascribed to the ε-constraint framework.

7.7 Conclusion

The resource allocation problem is a fundamental combinatorial optimization problem studied in the operations research community. The bi-objective uncertain stochastic resource allocation problem with the objectives of minimizing the risk of failure of completing activities and the resulting cost of resources is investigated in this chapter. A general mathematical formulation of the bi-objective uncertain stochastic resource allocation problem by using the CVaR measure to control the risk induced by uncertainties is proposed for the first time. Then based on the generalized models, formulations of two examples, i.e., RAPs and MWTA problems, are given. In order to determine referenced Pareto fronts, a linearized formulation of the RAP and an approximated lower bound formulation of the MWTA problem are presented according to problem-specific characteristics. A true PF and a lower bound PF can be obtained for the RAP and MWTA problem, respectively. Two state-of-the-art decomposition-based MOEAs (i.e., MOEA/D-AWA and DMOEA-εC) are used to solve the formulated problems. With the aim of having a fair comparison between MOEA/D-AWA and DMOEA-εC, two matching schemes with different matching criteria inspired by DMOEA-εC are proposed and embedded in MOEA/D-AWA.

Computational experiments have been performed on a set of the RAP and MWTA instances. Numerical experiments have demonstrated that both MOEA/D-AWA-M and DMOEA-εC are efficient ways of solving the bi-objective uncertain stochastic resource allocation problem and DMOEA-εC shows significant better performance than MOEA/D-AWA-M on RAP2 and the medium-scaled MWTA test instances. Besides, DMOEA-εC and MOEA/D-AWA-M perform similarly on RAP1 and small-scaled MWTA test instances. The superiority of DMOEA-εC can be ascribed to the ε-constraint framework.

Future research work includes investigating other types of uncertainties, adopting alternative methods to handle uncertainties, and embedding problem-specific knowledge in MOEAs to further improve their performance.

Epilogue

Most of the real-world problems encountered in practice are with multiple conflicting objective functions and affected by uncertainty. These problems are termed as uncertain multi-objective optimization problems (UMOPs). Multiple objectives and uncertainty are two important and difficult factors that should be taken into consideration in complex environments. Decomposition is basic strategy in the field of optimization. This book first reviews and summarizes the research results in the field of multi-objective optimization and uncertain optimization during recent years. When dealing with MOPs, this book reformulates MOPs into a series of scalar-constrained SOPs by incorporating the ε-constraint method into the decomposition strategy and thus proposes a decomposition-based multi-objective evolutionary algorithm with the ε-constraint framework (DMOEA-εC) to deal with MOPs. DMOEA-εC explicitly decomposes an MOP into a series of scalar-constrained optimization subproblems by selecting one of the objectives as the main objective function and assigning each subproblem with an upper bound vector. Then these subproblems are optimized simultaneously by evolving a population of solutions. At each generation, each individual solution in the population is associated with a subproblem. The neighborhood relations among these subproblems are defined based on the Euclidean distance between their upper bound vectors. The assumption that optimal solutions of two neighboring subproblems should be very similar is still valid. Besides, a main objective alternation strategy, a solution-to-subproblem matching procedure and a subproblem-to-solution matching procedure are proposed to strike a balance between convergence and diversity. A systematic experimental study has demonstrated that DMOEA-εC outperforms or performs competitively against other algorithms on the majority of the test instances.

In the literature, a category of MOPs that involve more than three objectives is termed as the MaOP. MaOPs appear widely in the field of industrial and engineering design. Many-objective optimization has been gaining attention in the evolutionary multi-objective optimization community during recent years. The high-dimensional objectives pose serious difficulties to MOEAs, which are originally designed for MOPs. In view of difficulties that DMOEA-εC is facing when dealing with MaOPs, a new efficient decomposition-based many-objective evolutionary algorithm with the ε-constraint framework, namely DMaOEA-εC, is proposed. Similar to DMOEA-εC, DMaOEA-εC explicitly decomposes an MOP into a series of scalar-constrained optimization subproblems by selecting one of the objectives as the main objective function and assigning each subproblem with an upper bound vector. When dealing with each scalar-constrained subproblem, a two-side update rule that maintains both feasible and infeasible solutions is proposed to make the update more effective, and thus enhance the convergence. Besides, in order to overcome the ineffectiveness induced by an exponential number of upper bound vectors, a two-stage upper bound vectors generation procedure is put forward to generate widely spread upper bound vectors in a high-dimensional space. Finally, a distance-based global replacement strategy and a boundary points maintenance mechanism are presented to remedy the diversity loss of a population and reduce the possibility of losing boundary solutions, respectively. DMaOEA-εC has been compared with several state-of-the-art and classical multi- and many-objective evolutionary algorithms to exhibit its performance on both MOPs and MaOPs. Experimental results confirm that DMaOEA-εC can successfully deal with the majority of test instances of MOPs and MaOPs.

As to UOPs, to our best knowledge, current existing techniques on dealing with UOPs are all *a priori* methods. Actually, it may be obtrusive or even risky to incorporate the preferences of a DM to handle uncertainties when he/she does not have sufficient knowledge about the problem. This book treats UOPs in an *a posteriori* manner and proposes a subproblems co-solving evolutionary algorithm, i.e., S-CoEA, to cope with UOPs. S-CoEA explicitly decomposes a UOP into a series of subproblems that represents different deterministic models by assigning each subproblem with a weight vector. These subproblems are formulated based on different aggregation forms of a sequence of sorted sampled function values and represent different preferences on uncertainties. Then, these subproblems are solved in parallel by evolving a population of solutions. At each generation, each individual solution in the population is associated with a subproblem. The neighborhood relations among these subproblems are defined based on the Euclidean distance between their weight vectors. Besides, a sample-updating strategy based on historical information is proposed for enhancing the performance of S-CoEA. S-CoEA has been compared with competitors on continuous test instances and RAPs with various characteristics and different strength levels of uncertainties. A systematical experimental study has shown that S-CoEA outperforms or performs competitively against comparison algorithms on the majority of test instances. Actually, the proposed algorithm is a decomposition-based *a posteriori* decision-making framework that can be extended in the future. For example, when considering the second-order moment, another criterion $V_f(\mathbf{x}, \alpha, \xi) = \sum_{i=1}^{n} r_{f,i} \cdot (F_w(\mathbf{x}, \alpha, \xi) - f(\mathbf{x}, \alpha, \xi_i))^2$ should be added, where $F_w(\mathbf{x}, \alpha, \xi) = \sum_{i=1}^{n} w_{f,i} \cdot f(\mathbf{x}, \alpha, \xi_i)$. $r_f = (r_{f,1}, ..., r_{f,n})(\sum_{i=1}^{n} r_{f,i} = 1)$ means the weight vector of the second-order moment. Similar results can be obtained for uncertain constraint functions $g_j(\mathbf{x}, \alpha, \xi)(j = 1, ..., p)$.

Uncertainties widely exist in real-world problems, which are mostly with multiple conflicting objectives. Usually the uncertainty is modeled as additive noise in the objective space. The decomposition-based multi-objective evolutionary algorithms (DMOEAs) are important approaches for multi-objective optimization and have not been applied to uncertain problems extensively. In order to design effective noise-tolerant DMOEAs (NT-DMOEAs) for uncertain problems, this book has conducted extensive studies to examine the impact of noise on DMOEAs, particularly for the population dynamics of convergence and diversity. Based on observations of noise impacts on DMOEAs, four noise-handling techniques have been proposed and embedded in two popular DMOEAs, i.e., MOEA/D and DMOEA-εC. The proposed NT-DMOEAs utilize the composite benefits of four major extensions. Two NT-DMOEAs have been compared with their various variants and 4 noise-tolerant algorithms to show the superiority of proposed features on 17 test instances with different strength levels of noise. A systematical experimental study has shown that two NT-DMOEAs significantly outperform their variants on the majority of test instances with different strength levels of noise. It also has shown that two NT-DMOEAs, especially the NT-DMOEA-εC, exhibit competitive or superior performance in terms of proximity and diversity for the majority of test instances.

The majority of practical problems are characterized with multiple objectives and uncertainty. For example, given a graph, the critical node detection problem consists of finding a set of nodes, the deletion of which satisfies one or more metrics in the induced graph. In contrast to most previous approaches, we use a bi-objective formulation, rather than make hypotheses on the psychology of decision makers. In this book, we propose and study a new variant of this problem called the bi-objective critical node detection problem (Bi-CNDP). In this formulation, we assume that removing each node has a cost, and decision makers want to minimize the pairwise connectivity of the induced graph and minimize the cost of removing a set of nodes at the same time. We firstly prove the NP-hardness of this problem on general graphs and the existence of a polynomial algorithm for constructing the ε-approximated Pareto front (PF) for Bi-CNDPs on trees. Then, different types of the mating pool and the replacement pool are proposed and integrated in two state-of-the-art decomposition-based MOEAs, including MOEA/D and DMOEA-εC. Two MOEAs and their variants are applied to solve the proposed Bi-CNDP. Sixteen common benchmark problems are modified by assigning random and logarithmic weight to each node. Computational experiments on all test instances were conducted first to evaluate the performance of different variants about the mating pool and the replacement pool. Then, further numerical experiment are used to compare the performance of two improved MOEAs, i.e., I-MOEA/D and I-DMOEA-εC, on Bi-CNDP. Numercial results not only show the effectivenesses of the proposed fourth mating pool and the global replacement strategy, but also demonstrate different behaviors of two improved MOEAs and the superiority of I-DMOEA-εC on the majority of test problems. Finally, a decision-making process from the perspective of a single-objective is proposed for helping decision makers to identify the most critical nodes with the smallest pairwise connectivity and the total cost of removing nodes that are less than a predefined threshold.

The resource allocation problem is a fundamental combinatorial optimization problem studied in the operations research community. The bi-objective uncertain stochastic resource allocation problem with the objectives of minimizing the risk of failure of completing activities and the resulting cost of resources is investigated in this book. A general mathematical formulation of the bi-objective uncertain stochastic resource allocation problem by using the CVaR measure to control the risk induced by uncertainties is proposed for the first time. Then based on the generalized models, formulations of two examples, i.e., RAPs and MWTA problems, are given. In order to determine referenced PFs, a linearized formulation of the RAP and an approximated lower bound formulation of the MWTA problem are presented, according to problem-specific characteristics. A true PF and a lower bound PF can be obtained for the RAP and MWTA problem, respectively. Two state-of-the-art decomposition-based multi-objective evolutionary algorithms (i.e., MOEA/D-AWA and DMOEA-εC) are used to solve the formulated problems. With the aim of having a fair comparison between MOEA/D-AWA and DMOEA-εC, two matching schemes with different matching criteria inspired by DMOEA-εC are proposed and embedded in MOEA/D-AWA. Computational experiments have been performed on a set of the RAP and MWTA instances. Numerical experiments have demonstrated that both MOEA/D-AWA-M and DMOEA-εC are efficient ways of solving the bi-objective uncertain stochastic resource allocation problem and DMOEA-εC shows significant better performance than MOEA/D-AWA-M on the majority of test instances.

For the DMOEA-εC framework, future research work includes investigations of adopting alternative methods to solve each constrained subproblem, employing more effective methods for estimating the nadir point and proposing an adjustment strategy for upper bound vectors to further improve the uniformity of the final population. Besides, based on our previous research works [Chen *et al.* (2009b); Xin *et al.* (2012)], the hybridization of different search operators in DMOEA-εC is also worth studying.

Many real-world problems suffer from noisy disturbances. Under such circumstances, standard DMOEAs are inadequate to provide Pareto optimal solutions for these noisy problems, which calls for the necessity of noise-tolerant algorithms. Actually, it is usually difficult to give a good representation of noise in real environments, thus estimating the noise accurately is an important task when handling noisy optimization problems. The most common way of estimating noise is to sample candidate solutions multiple times and measure the strength level of noise based on the variance value of a series of samples. This method is simple and straightforward, but not effective enough when the characteristics of noise are difficult to describe. What's more, there are many different forms of noise whose characteristics change as a function of location (in the design or objective space), or which alter during the course of optimization [Fieldsend and Everson (2015)]. This poses new challenges to the uncertain optimization. It will be our future work to propose an effective way of giving a good representation of the noise and estimating the strength level of noise accurately. Furthermore, new mechanisms that can deal with other types of uncertainties will be investigated. The performances of proposed NT-DMOEAs will also be tested on more complex real-world problems.

When applying a certain algorithm to solve a real-world problem, future research work will include investigations of designing more effective reproduction operators, embedding problem-specific knowledge during the optimization process, and using single-objective methods to further refine solutions obtained via multi-objective approaches. Besides, we can also consider investigating other types of uncertainties, adopt alternative methods for handling uncertainties, and embed problem-specific knowledge in multi-objective evolutionary algorithms to further improve their performance.

Bibliography

Addis, B., Aringhieri, R., Grosso, A., and Hosteins, P. (2016). Hybrid constructive heuristics for the critical node problem, *Annals of Operations Research* **238**, 1, pp. 1–13.

Addis, B., Di Summa, M., and Grosso, A. (2013). Identifying critical nodes in undirected graphs: Complexity results and polynomial algorithms for the case of bounded treewidth, *Discrete Applied Mathematics* **161**, 16–17, pp. 2349–2360.

Adra, S. F. and Fleming, P. J. (2011). Diversity management in evolutionary many-objective optimization, *IEEE Transactions on Evolutionary Computation* **15**, 2, pp. 183–195.

Ahuja, R. K., Kumar, A., Jha, K. C., and Orlin, J. B. (2007). Exact and heuristic algorithms for the weapon-target assignment problem, *Operations Research* **55**, 6, pp. 1136–1146.

Alves, M. J. and Costa, J. P. (2009). An exact method for computing the nadir values in multiple objective linear programming, *European Journal of Operational Research* **198**, 2, pp. 637–646.

Aringhieri, R., Grosso, A., Hosteins, P., and Scatamacchia, R. (2015). VNS solutions for the critical node problem, *Electronic Notes in Discrete Mathematics* **47**, pp. 37–44.

Aringhieri, R., Grosso, A., Hosteins, P., and Scatamacchia, R. (2016a). A general evolutionary framework for different classes of critical node problems, *Engineering Applications of Artificial Intelligence* **55**, pp. 128–145.

Aringhieri, R., Grosso, A., Hosteins, P., and Scatamacchia, R. (2016b). Local search metaheuristics for the critical node problem, *Networks* **67**, 3, pp. 209–221.

Aringhieri, R., Grosso, A., Hosteins, P., and Scatamacchia, R. (2018). Polynomial and pseudo-polynomial time algorithms for different classes of the distance critical node problem, *Discrete Applied Mathematics* **253**, pp. 103–121.

Arulselvan, A., Commander, C. W., Elefteriadou, L., and Pardalos, P. M. (2009). Detecting critical nodes in sparse graphs, *Computers and Operations Research* **36**, 7, pp. 2193–2200.

Arulselvan, A., Commander, C. W., Pardalos, P. M., and Shylo, O. (2007). Managing network risk via critical node identification, *Risk Management in Telecommunication Networks*, Springer.

Arulselvan, A., Commander, C. W., Shylo, O., Pardalos, P. M. (2011). *Cardinality-Constrained Critical Node Detection Problem.* In: Gülpınar, N., Harrison, P., Rüstem B. (eds.) Performance Models and Risk Management in Communications Systems. Springer Optimization and Its Applications, Vol. 46. Springer, New York, NY.

Aspnes, J., Chang, K., and Yampolskiy, A. (2005). Inoculation strategies for victims of viruses and the sum-of-squares partition problem, *Proceedings of the 16th Annual ACM-SIAM Symposium on Discrete Algorithms*, pp. 43–52.

Atputharajah, A. and Saha, T. K. (2009). Power system blackouts—literature review, *Proceedings of the International Conference on Industrial and Information Systems*, pp. 460–465.

Bader, J. and Zitzler, E. (2011). Hype: An algorithm for fast hypervolume-based many-objective optimization, *Evolutionary Computation* **19**, 1, pp. 45–76.

Balas, E. and Souza, C. C. d. (2005). The vertex separator problem: A polyhedral investigation, *Mathematical Programming* **103**, 3, pp. 583–608.

Bechikh, S. (2013). Incorporating decision maker's preference information in evolutionary multi-objective optimization, Ph.D. thesis, Tunis University, Tunis, Tunisia.

Bechikh, S., Said, L. B., and Ghédira, K. (2011). Searching for knee regions of the Pareto front using mobile reference points, *Soft Computing* **15**, 9, pp. 1807–1823.

Ben-Ameur, W., Mohamed-Sidi, M.-A., and Neto, J. (2015). The k-separator problem: Polyhedra, complexity and approximation results, *Journal of Combinatorial Optimization* **29**, 1, pp. 276–307.

Beyer, H. G., Olhofer, M., and Sendhoff, B. (2003). On the behavior of $(\mu/\mu_I, \lambda)$-ES optimizing functions disturbed by generalized noise, *Proceedings of the Foundations of the Genetic Algorithms*, pp. 307–328.

Beyer, H. G. and Sendhoff, B. (2006). Functions with noise-induced multimodality: A test for evolutionary robust optimization properties and performance analysis, *IEEE Transactions on Evolutionary Computation* **10**, 5, pp. 507–526.

Beyer, H. G. and Sendhoff, B. (2007). Robust optimization: A comprehensive survey, *Computer Methods in Applied Mechanics and Engineering* **196**, 33–34, pp. 3190–3218.

Blodgett, D. E., Gendreau, M., and Potvin, J. Y. (2003). A Tabu search heuristic for resource management in naval warfare, *Journal of Heuristics* **9**, 2, pp. 145–169.

Boonma, P. and Suzuki, J. (2009). A confidence-based dominance operator in evolutionary algorithms for noisy multiobjective optimization problems, *Proceedings of 21st IEEE International Conference on Tools with Artificial Intelligence*, pp. 387–394.

Borgatti, S. P. (2006). Identifying sets of key players in a social network, *Computational and Mathematical Organization Theory* **12**, 1, pp. 21–34.

Braford, J. C. (1961, AEREITM-9). Determination of optimal assignment of a weapon system to several targets, Technical report, Vought Aeronautics, Dallas, Texas.

Branke, J., Kaußler, T., and Schmeck, H. (2001). Guidance in evolutionary multi-objective optimization, *Advances in Engineering Software* **32**, 6, pp. 499–507.

Branke, J. and Schmidt, C. (2003). Selection in the presence of noise, *Proceedings of the Conference on Genetic and Evolutionary Computation*, pp. 766–777.

Brockhoff, D. and Zitzler, E. (2006). Are all objectives necessary? On dimensionality reduction in evolutionary multiobjective optimization, *Proceedings of the International Conference on Parallel Problem Solving From Nature*, pp. 533–542.

Buche, D., Stall, P., Dornberger, R., and Koumoutsakos, P. (2002). Multiobjective evolutionary algorithm for the optimization of noisy combustion processes, *IEEE Transactions on Systems, Man and Cybernetics, Part C* **32**, 4, pp. 460–473.

Bui, L., Abbass, H., and Essam, D. (2005). Fitness inheritance for noisy evolutionary multi-objective optimization, *Proceedings of the Genetic and Evolutionary Computation Conference*, pp. 779–785.

Cai, H., Liu, J., Chen, Y., and Wang, H. (2006). Survey of the research on dynamic weapon-target assignment problem, *Journal of Systems Engineering and Electronics* **17**, 3, pp. 559–565.

Cai, H. P. (2010). Study on firepower optimal distribution of artillery group based on NSGA-II, Ph.D. thesis, National University of Defense Technology, Changsha, China.

Cantú-Paz, E. (2004). Adaptive sampling for noisy problems, *Proceedings of Genetic and Evolutionary Computation Conference* (Springer, Berlin Heidelberg), pp. 947–958.

Caserta, M. and Voß, S. (2016). A corridor method based hybrid algorithm for redundancy allocation, *Journal of Heuristics* **22**, 4, pp. 405–429.

Chen, B., Zeng, W., and Lin, Y. (2015). A new local search-based multiobjective optimization algorithm, *IEEE Transactions on Evolutionary Computation* **19**, 1, pp. 50–73.

Chen, C., Chen, C., and Zhang, Q. (2009a). Enhancing MOEA/D with guided mutation and priority update for multi-objective optimization, *Proceedings of the IEEE Congress on Evolutionary Computation*, pp. 209–216.

Chen, C. H. (1995). An effective approach to smartly allocate computing budget for discrete event simulation, *Proceedings of the 34th IEEE Conference on Decision and Control*, pp. 2598–2605.

Chen, C. H., Wu, S. D., and Dai, L. (1999). Ordinal comparison of heuristic algorithms using stochastic optimization, *IEEE Transactions on Robotics and Automation* **15**, 1, pp. 44–56.

Chen, J., Li, J., and Xin, B. (2017). DMOEA-εC: Decomposition-based multi-objective evolutionary algorithm with the ε-constraint framework, *IEEE Transactions on Evolutionary Computation* **21**, 5, pp. 714–730.

Chen, J., Xin, B., Peng, Z. H., Dou, L. H., and Zhang, J. (2009b). Optimal contraction theorem for exploration-exploitation trade-off in search and optimization, *IEEE Transactions on Systems, Man and Cybernetics, Part A: Systems and Humans* **39**, 3, pp. 680–691.

Chen, P.-A., David, M., and Kempe, D. (2010). Better vaccination strategies for better people, *Proceedings of the ACM Conference on Electronic Commerce*, pp. 179–188.

Chen, T. C. and You, P. S. (2005). Immune algorithms-based approach for redundant reliability problems with multiple component choices, *Computers in Industry* **56**, pp. 195–205.

Chern, M. (1992). On the computational complexity of reliability redundancy allocation in a series system, *Operations Research Letters* **11**, 5, pp. 309–315.

Chu, H. Y., Xu, P. P., and Yang, C. C. (2016). Joint relay selection and power control for robust cooperative multicast in mmWave WPANs, *Science China-Information Sciences* **59**, 8, p. No. 082301.

Coello, C. A. C. (2009). Study of preference relations in many-objective optimization, *Proceedings of the Conference on Genetic and Evolutionary Computation*, pp. 611–618.

Coello, C. A. C. and Chakraborty, D. (2008). Objective reduction using a feature selection technique, *Proceedings of the Conference on Genetic and Evolutionary Computation*, pp. 673–680.

Cvetkovic, D. and Parmee, I. C. (2002). Preferences and their application in evolutionary multiobjective optimization, *IEEE Transactions on Evolutionary Computation* **6**, 1, pp. 42–57.

Das, I. and Dennis, J. E. (1998). Normal-boundary intersection: A new method for generating the pareto surface in nonlinear multicriteria optimization, *SIAM Journal on Optimization* **8**, 3, pp. 631–657.

Day, R. H. (1966). Allocating weapons to target complexes by means of nonlinear programming, *Operations Research* **14**, 6, pp. 992–1013.

Deb, K. (2000). An efficient constraint handling method for genetic algorithms, *Computer Methods in Applied Mechanics and Engineering* **186**, 2–4, pp. 311–338.

Deb, K., Agrawal, S., Pratap, A., and Meyarivan, T. (2002a). A fast and elitist multiobjective genetic algorithm: NSGA-II, *IEEE Transactions on Evolutionary Computation* **6**, 2, pp. 182–197.

Deb, K., J. Sundar, N. U., and Chaudhuri, S. (2006). Reference point based multi-objective optimization using evolutionary algorithms, *International Journal of Computational Intelligence Research* **2**, 6, pp. 273–286.

Deb, K. and Jain, H. (2014). An evolutionary many-objective optimization algorithm using reference-point-based nondominated sorting approach, Part I: solving problems with box constraints, *IEEE Transactions on Evolutionary Computation* **18**, 4, pp. 577–601.

Deb, K. and Saxena, D. K. (2005). On finding Pareto-optimal solutions through dimensionality reduction for certain large-dimensional multi-objective optimization problems, Kangal Report, Indian Institute of Technology, Kharagpur, India.

Deb, K. and Saxena, D. K. (2006). Searching for Pareto-optimal solutions through dimensionality reduction for certain large-dimensional multiobjective optimization problems, *Proceedings of the IEEE Congress on Evolutionary Computation*, pp. 3353–3360.

Deb, K., Thiele, L., Laumanns, M., and Zitzler, E. (2002b). Scalable multiobjective optimization test problems, *Proceedings of the IEEE Congress on Evolutionary Computation*, pp. 825–830.

denBroeder, G., Ellison, R., and Emerling, L. (1959). On optimum target assignments, *Operations Research* **7**, 3, pp. 322–326.

Di Summa, M., Grosso, A., and Locatelli, M. (2011). Complexity of the critical node problem over trees, *Computers and Operations Research* **38**, 12, pp. 1766–1774.

Di Summa, M., Grosso, A., and Locatelli, M. (2012). Branch and cut algorithms for detecting critical nodes in undirected graphs, *Computational Optimization and Applications* **53**, 3, pp. 649–680.

Dinh, T. N. and Thai, M. T. (2013). Precise structural vulnerability assessment via mathematical programming, *Proceedings of the Military Communications Conference*, pp. 1351–1356.

Dinh, T. N., Xuan, Y., Thai, M. T., Park, E. K., and Znati, T. (2010). On approximation of new optimization methods for assessing network vulnerability, *Proceedings of the Conference on Information Communications*, pp. 2678–2686.

Durillo, J. J. and Nebro, A. J. (2011). jMetal: A JAVA framework for developing multi-objective optimization, *Advances in Engineering Software* **42**, 10, pp. 760–771.

Esquivel, X., Esquivel, X., Lara, A., and Coello, C. A. C. (2012). Using the averaged Hausdorff distance as a performance measure in evolutionary multiobjective optimization, *IEEE Transactions on Evolutionary Computation* **16**, 4, pp. 504–522.

Fan, N. and Pardalos, P. M. (2010). *Robust Optimization of Graph Partitioning and Critical Node Detection in Analyzing Networks.* In: Wu, W. and Daescu, O. (eds.) Combinatorial Optimization and Applications. COCOA 2010. Lecture Notes in Computer Science, Vol. 6508. Springer, Berlin, Heidelberg.

Faramondi, L., Oliva, G., Panzieri, S., Pascucci, F., Schlueter, M., Munetomo, M., and Setola, R. (2018). Network structural vulnerability: A multiobjective attacker perspective, *IEEE Transactions on Systems, Man and Cybernetics Systems* **PP**, 99, pp. 1–14.

Farina, M. and Amato, P. (2004). A fuzzy definition of optimality for many-criteria optimization problems, *IEEE Transactions on Systems, Man and Cybernetics, Part A: Systems and Humans* **34**, 3, pp. 315–326.

Fieldsend, J. E. and Everson, R. M. (2005). Multi-objective optimisation in the presence of uncertainty evolutionary computation, *Proceedings of the IEEE Congress on Evolutionary Computation*, pp. 243–250.

Fieldsend, J. E. and Everson, R. M. (2015). The rolling tide evolutionary algorithm: A multiobjective optimizer for noisy optimization problems, *IEEE Transactions on Evolutionary Computation* **19**, 1, pp. 103–117.

Fonseca, C. and Fleming, P. (1993). Genetic algorithm for multiobjective optimization: Formulation, discussion and generation, *Proceedings of the International Conference on Genetic Algorithms*, pp. 416–423.

Giagkiozis, I., Purshouse, R. C., and Fleming, P. J. (2013). Generalized decomposition, in *Evolutionary Multi-Criterion Optimization* (Springer, Berlin, Heidelberg), pp. 428–442.

Goh, C. K. and Tan, K. C. (2007). An investigation on noisy environments in evolutionary multiobjective optimization, *IEEE Transactions on Evolutionary Computation* **11**, 3, pp. 354–381.

Gonçalves, R. A., Almeida, C. P., and Pozo, A. (2015). *Upper Confidence Bound (UCB) Algorithms for Adaptive Operator Selection in MOEA/D* (Springer, Berlin).

Grandinetti, L., Pisacane, O., and Sheikhalishahi, M. (2013). An approximate ε-constraint method for a multi-objective job scheduling in the cloud, *Future Generation Computer Systems* **29**, 8, pp. 1901–1908.

Gu, F. and Liu, H. (2010). A novel weight design in multi-objective evolutionary algorithm, *Proceedings of the International Conference on Computational Intelligence and Security*, pp. 137–141.

Hadka, D. and Reed, P. (2012). Diagnostic assessment of search controls and failure modes in many-objective evolutionary optimization, *Evolutionary Computation* **20**, 3, pp. 423–452.

Haimes, Y. Y., Lasdon, S. S., and Wismer, D. A. (1971). On a bicriterion formulation of the problems of integrated system identification and system optimization, *IEEE Transactions on Systems, Man and Cybernetics* **1**, 3, pp. 296–297.

Hao, J. K., Dorne, R., and Galinier, P. (1998). Tabu search for frequency assignment in mobile radio networks, *Journal of Heuristics* **4**, 1, pp. 47–62.

Hernandez Gomez, R. and Coello Coello, C. (2013). Mombi: A new metaheuristic for many-objective optimization based on the r2 indicator, *Proceedings of the IEEE Congress on Evolutionary Computation*, pp. 2488–2495.

Horn, J., Nafpliotis, N., and Goldberg, D. (1994). A niched Pareto genetic algorithm for multiobjective optimization, *Proceedings of the IEEE Congress on Evolutionary Computation*, pp. 82–87.

Hosein, P. A. and Athans, M. (1990a). Preferential defense strategies. Part II: The dynamic case, Technical report, Laboratory for Information & Decision Systems, Cambridge, United Kingdom.

Hosein, P. A. and Athans, M. (1990b). Some analytical results for the dynamic weapon-target allocation problem, Technical report, MIT, Cambridge, United Kingdom.

Hosein, P. A., Walton, J. T., and Athans, M. (1988). Dynamic weapon-target assignment problems with vulnerable C2 nodes, Technical report, Laboratory for Information & Decision Systems, Cambridge, United Kingdom.

Huang, C. L. (2015). A particle-based simplified swarm optimization algorithm for reliability redundancy allocation problems, *Reliability Engineering and System Safety* **142**, pp. 221–230.

Huang, W. and Li, H. (2010). On the differential evolution schemes in MOEA/D, *Proceedings of the International Conference on Natural Computation*, pp. 2788–2792.

Huang, X. (2007). Portfolio selection with fuzzy returns, *Journal of Intelligent and Fuzzy Systems Applications in Engineering and Technology* **18**, 4, pp. 383–390.

Huang, X. (2008). Mean-semivariance models for fuzzy portfolio selection, *Journal of Computational and Applied Mathematics* **16**, 4, pp. 1096–1101.

Huband, S., Barone, L., While, L., and Hingston, P. (2005). A scalable multiobjective optimization test problem toolkit, *Proceedings of the 3rd International Conference on Evolutionary Multi-Criterion Optimization*, pp. 280–295.

Huband, S., Hingston, P., While, L., and Barone, L. (2003). An evolution strategy with probabilistic mutation for multi-objective optimisation, *Proceedings of the IEEE Congress on Evolutionary Computation*, pp. 2284–2291.

Hughes, E. J. (2001). Evolutionary multi-objective ranking with uncertainty and noise, *Proceedings of International Conference on Evolutionary Multi-Criterion Optimization*, pp. 329–343.

Hughes, E. J. (2003). Multiple single objective Pareto sampling, *Proceedings of the IEEE Congress on Evolutionary Computation*, Vol. 4, pp. 2678–2684.

Hwang, C. L. and Masud, A. S. M. (1979). *Multiple Objectives Decision Making Methods and Applications* (Springer, Berlin).

Ishibuchi, H., Sakane, Y., Tsukamoto, N., and Nojima, Y. (2009). Adaptation of scalarizing functions in MOEA/D: An adaptive scalarizing function-based multiobjective evolutionary algorithm, *Proceedings of the International Conference on Evolutionary Multi-Criterion Optimization*, pp. 438–452.

Ishibuchi, H., Sakane, Y., Tsukamoto, N., and Nojima, Y. (2010). Simultaneous use of different scalarizing functions in MOEA/D, *Proceedings of the Genetic and Evolutionary Computation Conference*, pp. 519–526.

Ishibuchi, H., Tanigaki, Y., Masuda, H., and Nojima, Y. (2014). *Distance-Based Analysis of Crossover Operators for Many-Objective Knapsack Problems* (Springer International Publishing).

Ishibuchi, H., Tsukamoto, N., and Nojima, Y. (2008). Behavior of evolutionary many-objective optimization, *Proceedings of the 10th International Conference on Computer Modeling and Simulation*, pp. 266–271.

Ishibuchi, H., Yu, S., Masuda, H., and Nojima, Y. (2016). Performance of decomposition-based many-objective algorithms strongly depends on Pareto front shapes, *IEEE Transactions on Evolutionary Computation* **21**, 2, pp. 169–190.

Jaimes, A. L., Coello, C. A. C., Aguirre, H., and Tanaka, K. (2011). *Adaptive Objective Space Partitioning Using Conflict Information for Many-Objective Optimization*. In: Takahashi, R. H. C., Deb, K., Wanner, E. F. and Greco, S. (eds.) Evolutionary Multi-Criterion Optimization. EMO 2011. Lecture Notes in Computer Science, Vol. 6576. Springer, Berlin, Heidelberg.

Jaszkiewicz, A. (2002). On the performance of multi-objective genetic local search on 0/1 knapsack problem—A comparative experiment, *IEEE Transactions on Evolutionary Computation* **6**, 4, pp. 402–412.

Jenelius, E., Petersen, T., and Mattsson, L.-G. (2006). Importance and exposure in road network vulnerability analysis, *Transportation Research, Part A: Policy and Practice* **40**, 7, pp. 537–560.

Ji, Z. G., Chen, H. B., and Li, X. Y. (2019). Design for reliability with the advanced integrated circuit (ic) technology: challenges and opportunities, *SCIENCE CHINA Information Sciences*, Vol. 62, no. 12. pp. 226401: 1–226401: 4.

Jiang, C., Han, X., and Liu, G. P. (2008). A nonlinear interval number programming method for uncertain optimization problems, *European Journal of Operational Research* **188**, 1, pp. 1–13.

Jiang, S. W., Cai, Z. H., Zhang, J., and Ong, Y. S. (2011). Multiobjective optimization by decomposition with Pareto-adaptive weight vectors, *Proceedings IEEE International Conference on Natural Computation*, pp. 1260–1264.

Jin, Y., Olhofer, M., and Sendho, B. (2001). Dynamic weighted aggregation for evolutionary multi-objective optimization: Why does it work and how? *Genetic and Evolutionary Computation Conference.*

Jin, Y. C. and Branke, J. (2005). Evolutionary optimization in uncertain environments—A survey, *IEEE Transactions on Evolutionary Computation* **9**, 3, pp. 303–317.

Johnson, B. L., Porter, A. T., King, J. C., and Newman, A. M. (2018). Optimally configuring a measurement system to detect diversions from a nuclear fuel cycle, *Annals of Operations Research.*

Kasprzyk, J. R., Reed, P. M., Kirsch, B. R., and Characklis, G. W. (2009). Managing population and drought risks using many-objective water portfolio planning under uncertainty, *Water Resources Research* **45**, 12, pp. 170–170.

Kaushik, B., Kaur, N., and Kohli, A. K. (2013). Achieving maximum reliability in fault tolerant network design for variable networks, *Applied Soft Computing* **13**, 7, pp. 3211–3224.

Ke, L., Zhang, Q., and Battiti, R. (2013). MOEA/D-ACO: A multiobjective evolutionary algorithm using decomposition and ant colony, *IEEE Transactions on Cybernetics* **43**, 6, pp. 1845–1859.

Kempe, D., Kleinberg, J., and Tardos, E. (2010). Maximizing the spread of influence in a social network, *Progressive Research*, pp. 137–146.

Khalili-Damghani, K. and Amiri, M. (2012). Solving binary-state multi-objective reliability redundancy allocation series-parallel problem using efficient epsilon-constraint, multi-start partial bound enumeration algorithm, and DEA, *Reliability Engineering and System Safety* **103**, 4, pp. 35–44.

Kline, A., Ahner, D., and Hill, R. (2019). The weapon-target assignment problem, *Computers and Operations Research* **105**, pp. 226–236.

Knowles, J., Corne, D., and Reynolds, A. (2009). Noisy multiobjective optimization on a budget of 250 evaluations, *Proceedings of International Conference on Evolutionary Multi-Criterion Optimization*, pp. 36–50.

Kruisselbrink, J. W., Emmerich, M. T., Bender, A., Ijzerman, A. P., and Horst, E. (2009). Combining aggregation with Pareto optimization: A case study in evolutionary molecular design, *Proceedings of the International Conference on Evolutionary Multi-Criterion Optimization*, pp. 453–467.

Kuhlman, C. J., Kumar, V. S. A., Marathe, M. V., Ravi, S. S., and Rosenkrantz, D. J. (2010). Finding critical nodes for inhibiting diffusion of complex contagions in social networks, *Proceedings of the European Conference on Machine Learning and Knowledge Discovery in Databases*, pp. 111–127.

Kukkonen, S. and Deb, K. (2006). A fast and effective method for pruning of non-dominated solutions in many-objective problems, *Proceedings of International Conference on Parallel Problem Solving From Nature*, pp. 553–562.

Kumar, V. S. A., Rajaraman, R., Sun, Z., and Sundaram, R. (2010). Existence theorems and approximation algorithms for generalized network security games, *Proceedings of the IEEE International Conference on Distributed Computing Systems*, pp. 348–357.

Lai, Y. (2009). Multiobjective optimization using MOEA/D with a new mating selection mechanism, Master's thesis, Taiwan Normal University.

Laili, Y., Zhang, L., Tao, F., and Ma, P. (2016). Rotated neighbor learning-based auto-configured evolutionary algorithm, *Science China Information Sciences* **53**, 5, pp. 150–163.

Lalou, M., Tahraoui, M. A., and Kheddouci, H. (2016). Component-cardinality-constrained critical node problem in graphs, *Discrete Applied Mathematics* **210**, pp. 150–163.

Lalou, M., Tahraoui, M. A., and Kheddouci, H. (2018). The critical node detection problem in networks: A survey, *Computer Science Review* **28**, pp. 92–117.

Leboucher, C., Shin, H.-S., Siarry, P., Chelouah, R., Menec, S. L., and Tsourdos, A. (2013). A two-step optimisation method for dynamic weapon target assignment problem, Technical report, IntechOpen.

Lee, C. Y., Gen, M., and Kuo, W. (2002). Reliability optimization design using hybridized genetic algorithm with a neural network technique, *IEICE Transactions on Fundamentals of Electronics, Communications and Computer Sciences* **E84-A**, pp. 627–637.

Lee, M.-Z. (2010). Constrained weapon-target assignment: Enhanced very large scale neighborhood search algorithm, *Transactions on Systems, Man and Cybernetics Systems, Part A* **40**, 1, pp. 198–204.

Leskovec, J., Krause, A., Guestrin, C., Faloutsos, C., Vanbriesen, J., and Glance, N. (2007). Cost-effective outbreak detection in networks, *ACM SIGKDD International Conference on Knowledge Discovery and Data Mining*, pp. 420–429.

Li, H. and Landa-Silva, D. (2011). An adaptive evolutionary multi-objective approach based on simulated annealing, *Evolutionary Computation* **19**, 4, pp. 561–595.

Li, H. and Zhang, Q. (2009). Multiobjective optimization problems with complicated Pareto sets, MOEA/D and NSGA-II, *IEEE Transactions on Evolutionary Computation* **13**, 2, pp. 284–302.

Li, J., Chen, J., Xin, B., and Dou, L. H. (2015a). Solving multi-objective multi-stage weapon target assignment problem via adaptive NSGA-II and adaptive MOEA/D: A comparison study, *Proceedings of the IEEE Congress on Evolutionary Computation*, pp. 3132–3139.

Li, J., Chen, J., Xin, B., Dou, L. H., and Peng, Z. H. (2016a). Solving the uncertain multi-objective multi-stage weapon target assignment problem via MOEA/D-AWA, *Proceedings of IEEE Congress on Evolutionary Computation*, pp. 4934–4941.

Li, K., Deb, K., Zhang, Q. F., and Kwong, S. (2015b). An evolutionary many-objective optimization algorithm based on dominance and decomposition, *IEEE Transactions on Evolutionary Computation* **19**, 5, pp. 694–716.

Li, K., Zhang, Q., and Battiti, R. (2010a). Multiobjective combinatorial optimization by using decomposition and ant colony, Technical report, University of Essex, Colchestor, United Kingdom.

Li, M., Yang, S., and Liu, X. (2014). Shift-based density estimation for Pareto-based algorithms in many-objective optimization, *IEEE Transactions on Evolutionary Computation* **18**, 3, pp. 348–365.

Li, M. H., Huang, G. M., and Zhou, D. (2016b). A yield-enhanced global optimization methodology for analog circuit based on extreme value theory, *Science China Information Sciences* **59**, 8, p. No.082401.

Li, X., Qin, Z., and Kar, S. (2010b). Mean-variance-skewness model for portfolio selection with fuzzy returns, *European Journal of Operational Research* **202**, 1, pp. 239–247.

Li, Z. J., Liao, H. T., and Coit, D. W. (2009). A two-stage approach for multi-objective decision making with applications to system reliability optimization, *Reliability Engineering and System Safety* **94**, 10, pp. 1585–1592.

Liang, G., Weller, S. R., Zhao, J., Luo, F., and Dong, Z. Y. (2017). The 2015 Ukraine blackout: Implications for false data injection attacks, *IEEE Transactions on Power Systems* **32**, 4, pp. 3317–3318.

Liefooghe, A., Basseur, M., Jourdan, L., and Talbi, E. G. (2007). Combinatorial optimization of stochastic multi-objective problems: An application to the flow-shop scheduling problem, *Proceedings of IEEE International Conference on Evolutionary Multi-criterion Optimization*, pp. 457–471.

Liu, B. D. (2009). *Theory and Practice of Uncertain Programming* (Springer, Berlin, Heidelberg).

Liu, H. L., Gu, F., and Zhang, Q. (2014). Decomposition of a multi-objective optimization problem into a number of simple multi-objective subproblems, *IEEE Transactions on Evolutionary Computation* **18**, 3, pp. 450–455.

Liu, Y., Gong, D., Sun, J., and Jin, Y. (2017). A many-objective evolutionary algorithm using a one-by-one selection strategy, *IEEE Transactions on Cybernetics* **47**, 9, pp. 2689–2702.

Lloyd, S. P. and Witsenhausen, H. S. (1986). Weapons allocation is NP-complete, *Proceedings of IEEE Summer Conference on Simulation*, pp. 1054–1058.

Lu, R., Guan, X., Li, X., and Hwang, I. (2016). A large-scale flight multi-objective assignment approach based on multi-island parallel evolution algorithm with cooperative coevolutionary, *Science China Information Sciences* **59**, 7, pp. 1–17.

Lust, T. and Teghem, J. (2010). *The Multiobjective Traveling Salesman Problem: A Survey and a New Approach* (Springer, Berlin, Heidelberg).

Lygoe, R. J., Cary, M., and Fleming, P. J. (2013). *A Real-World Application of a Many-Objective Optimisation Complexity Reduction Process* (Springer, Berlin, Heidelberg).

Ma, X. L., Qi, Y. T., Li, L. L., Liu, F., Jiao, L. C., and Wu, J. S. (2014). MOEA/D with uniform decomposition measurement for many-objective problems, *Soft Computing* **18**, 12, pp. 1–24.

Manne, A. S. (1958). A target-assignment problem, *Operations Research* **6**, 3, pp. 346–351.

Matlin, S. (1970). A review of the literature on the missile allocation problem, *Operations Research* **18**, 2, pp. 334–373.

Mavrotas, G. (2009). Effective implementation of the ε-constraint method in multi-objective mathematical programming problems, *Applied Mathematics and Computation* **213**, 2, pp. 455–465.

Mavrotas, G. and Florios, K. (2013). An improved version of the augmented ε-constraint method (AUGMECON2) for finding the exact Pareto set in multi-objective integer programming problems, *Applied Mathematics and Computation* **219**, 18, pp. 9652–9669.

Mei, Y., Tang, K., and Yao, X. (2011). Decomposition-based memetic algorithm for multi-objective capacitated arc routing problem, *IEEE Transactions on Evolutionary Computation* **15**, 2, pp. 151–165.

Metev, B. and Vassilev, V. (2003). A method for nadir point estimation in MOLP problems, *Cybernetics and Information Technologies* **3**, 2, pp. 15–24.

Miettinen, K. (1999). *Nonlinear Multiobjective Optimization* (Kluwer Academic Publishers, Boston).

Mitchell, D. P. (1991). Spectrally optimal sampling for distribution ray tracing, *ACM Siggraph Computer Graphics* **25**, 4, pp. 157–164.

Montgomery, D. C. (2005). *Design and Analysis of Experiments* (John Wiley and Sons Press, Arizona).

Moubayed, N., Petrovski, A., and McCall, J. (2010). A novel multi-objective particle swarm optimizaiton based on decomposition, *Proceedings of the International Conference on Parallel Problem Solving from Nature*, pp. 1–10.

Moubayed, N. A., Petrovski, A., and Mccall, J. (2014). D(2)MOPSO: MOPSO based on decomposition and dominance with archiving using crowding distance in objective and solution spaces. *Evolutionary Computation* **22**, 1, pp. 47–77.

Nahas, N. and Nourelfath, M. (2005). Ant system for reliability optimization of a series system with multiple-choice and budget constraints, *Reliability Engineering and System Safety* **87**, pp. 1–12.

Nebro, A. J., Durillo, J. J., Luna, F., Dorronsoro, B., and Alba, E. (2009a). Mocell: A cellular genetic algorithm for multiobjective optimization, *International Journal of Intelligent Systems* **24**, 7, pp. 726–746.

Nebro, A. J., Durillo, J. J., Nieto, J. G., Coello, C. A. C., Luna, F., and Alba, E. (2009b). SMPSO: A new PSO-based metaheuristic for multi-objective optimization, *Proceedings of the IEEE Symposium on Computational Intelligence in Multi-criteria Decision-making*, pp. 66–73.

Nissen, V. and Propach, J. (1998). On the robustness of population based versus point-based optimization in the presence of noise, *IEEE Transactions on Evolutionary Computation* **2**, 3, pp. 107–119.

Papadimitriou, C. H. and Yannakakis, M. (2000). On the approximability of trade-offs and optimal access of web sources, *Proceedings of the Symposium on Foundations of Computer Science*, pp. 86–92.

Park, T. and Ryu, K. (2011). Accumulative sampling for noisy evolutionary multi-objective optimization, *Proceedings of the Genetic and Evolutionary Computation Conference*, pp. 793–800.

Pavlikov, K. (2018). Improved formulations for minimum connectivity network interdiction problems, *Computers and Operations Research* **97**, pp. 48–57.

Pierro, F. D., Khu, S. T., and Savic, D. A. (2007). An investigation on preference order ranking scheme for multiobjective evolutionary optimization, *IEEE Transactions on Evolutionary Computation* **11**, 1, pp. 17–45.

Praditwong, K. and Yao, X. (2006). A new multi-objective evolutionary optimisation algorithm: The two-archive algorithm, *Proceedings of the International Conference on Computational Intelligence and Security*, pp. 286–291.

Prasad, V. R. and Kuo, W. (2000). Reliability optimization of coherent systems, *IEEE Transactions on Reliability* **49**, pp. 323–330.

Purevsuren, D., Cui, G., Qu, M., and Win, N. N. H. (2017). Hybridization of GRASP with exterior path relinking for identifying critical nodes in graphs, *International Journal of Computer Science* **44**, 2, pp. 157–165.

Purshouse, R. C. and Fleming, P. J. (2007). On the evolutionary optimization of many conflicting objectives, *IEEE Transactions on Evolutionary Computation* **11**, 6, pp. 770–784.

Qi, Y., Ma, X., and Liu, F. (2014). MOEA/D with adaptive weight adjustment, *Evolutionary Computation* **22**, 2, pp. 231–264.

Rakshit, P. and Konar, A. (2015a). Differential evolution for noisy multiobjective optimization, *Artificial Intelligence* **227**, pp. 165–189.

Rakshit, P. and Konar, A. (2015b). Extending multi-objective differential evolution for optimization in presence of noise, *Information Sciences* **305**, pp. 56–76.

Rakshit, P., Konar, A., Das, S., Jain, L., and Nagar, A. (2014). Uncertainty management in differential evolution induced multi-objective optimization in presence of measurement noise, *IEEE Transactions on Systems, Man and Cybernetics Systems* **44**, 7, pp. 922–937.

Rockafellar, R. T. and Uryasev, S. (2000). Optimization of conditional value-at-risk, *Journal of Risk* **2**, 3, pp. 21–41.

Roy, P. C., Islam, M. M., Murase, K., and Yao, X. (2015). Evolutionary path control strategy for solving many-objective optimization problem, *IEEE Transactions on Cybernetics* **45**, 4, pp. 702–715.

Roy, R. K. (2010). *A Primer on the Taguchi Method* (Society of Manufacturing Engineers).

Said, L. B., Bechikh, S., and Ghedira, K. (2010). The r-dominance: A new dominance relation for interactive evolutionary multicriteria decision making, *IEEE Transactions on Evolutionary Computation* **14**, 5, pp. 801–818.

Salmeron, J., Wood, K., and Baldick, R. (2015). Analysis of electric grid security under terrorist threat, *IEEE Transactions on Power Systems* **19**, 2, pp. 905–912.

Saxena, D. K. and Deb, K. (2007). *Non-linear Dimensionality Reduction Procedures for Certain Large-dimensional Multi-objective Optimization Problems: Employing Correntropy and a Novel Maximum Variance Unfolding* (Springer, Berlin, Heidelberg).

Saxena, D. K., Duro, J. A., Tiwari, A., Deb, K., and Zhang, Q. (2013). Objective reduction in many-objective optimization: Linear and nonlinear algorithms, *IEEE Transactions on Evolutionary Computation* **17**, 1, pp. 77–99.

Schaffer, J. D. (1984). Some experiments in machine learning using vector evaluated genetic algorithms, Ph.D. thesis, Vanderbilt University, Nashville, Tennessee, United States.

Schaffer, J. D. (1985). Multiple objective optimization with vector evaluated genetic algorithms, *Proceedings of the 1st International Conference on Genetic Algorithms*.

Schervish, M. J. (2012). *Theory of Statistics* (Springer Science and Business Media, Berlin, Heidelberg).

Schutze, O., Esquivel, X., Lara, A., and Coello, C. A. C. (2012). Using the averaged Hausdorff distance as a performance measure in evolutionary multiobjective optimization, *IEEE Transactions on Evolutionary Computation* **16**, 4, pp. 504–522.

Sendhoff, B., Beyer, H. G., and Olhofer, M. (2002). On noise induced multimodality in evolutionary algorithms, *Proceedings of the Asia-Pacific Conference on Simulated Evolution and Learning*, pp. 219–224.

Shen, S. and Smith, J. C. (2012). Polynomial-time algorithms for solving a class of critical node problems on trees and series-parallel graphs, *Networks* **60**, 2, pp. 103–119.

Shen, Y., Dinh, T. N., and Thai, M. T. (2013a). Adaptive algorithms for detecting critical links and nodes in dynamic networks, *Proceedings of the IEEE Military Communications Conference*, pp. 1–6.

Shen, Y., Nguyen, N. P., Xuan, Y., and Thai, M. T. (2013b). On the discovery of critical links and nodes for assessing network vulnerability, *IEEE/ACM Transactions on Networking* **21**, 3, pp. 963–973.

Shim, V. A., Tan, K. C., and Tan, K. K. (2012). A hybrid estimation of distribution algorithm for solving the multi-objective multiple traveling salesman problem, *Proceedings of the IEEE Congress on Evolutionary Computation*, pp. 682–691.

Sindhya, K., Ruuska, S., Haanpaa, T., and Miettinen, K. (2011). A new hybrid mutation operator for multiobjective optimization with differential evolution, *Soft Computing* **15**, 10, pp. 2041–2055.

Singh, A. (2003). Uncertainty based multi-objective optimization of groundwater remediation design, Master's thesis, University Illinois at Urbana-Champaign, Urbana, Illinois, United States.

Singh, A. and Minsker, B. S. (2008). Uncertainty based multi-objective optimization of groundwater remediation design, *Water Resource Research* **440**, 2, pp. 1 –10.

Singh, H. K., Isaacs, A., and Ray, T. (2011). A Pareto corner search evolutionary algorithm and dimensionality reduction in many-objective optimization problems, *IEEE Transactions on Evolutionary Computation* **15**, 4, pp. 539–556.

Siwik, L. and Natanek, S. (2008). Elitist evolutionary multi-agent system in solving noisy multi-objective optimization problems, *Proceedings of IEEE Congress on Evolutionary Computation*, pp. 3319–3326.

Spears, W. M. and Jong, K. A. D. (1991). On the virtues of parameterized uniform crossover, *Proceedings of the 4th International Conference on Genetic Algorithms*, pp. 230–236.

Sulieman, D., Jourdan, L., and Talbi, E. (2010). Using multiobjective metaheuristics to solve VRP with uncertain demands, *Proceedings of IEEE International Conference on Evolutionary Computation*, pp. 1–8.

Sung, C. S. and Cho, Y. K. (2000). Reliability optimization of a series system with multiple-choice and budget constraints, *European Journal of Operational Research* **127**, pp. 158–171.

Syberfeldt, A., Ng, A., John, R., and Moore, P. (2010). Evolutionary optimization of noisy multi-objective problems using confidence-based dynamic resampling, *European Journal of Operational Research* **204**, 3, pp. 533–544.

Sülflow, A., Drechsler, N., and Drechsler, R. (2007). Robust multi-objective optimization in high dimensional spaces, *Proceedings of the International Conference on Evolutionary Multi-Criterion Optimization*, pp. 715–726.

Taboada, H. A., Baheranwala, F., and Coit, D. W. (2007). Practical solutions for multi-objective optimization: An application to system reliability design problems, *Reliability Engineering and System Safety* **92**, 3, pp. 314–332.

Tanaka, K. (2009). Space partitioning with adaptive ε-ranking and substitute distance assignments: A comparative study on many-objective mnk-landscapes, *Proceedings of the Conference on Genetic and Evolutionary Computation*, pp. 547–554.

Teich, J. (2001). Pareto-front exploration with uncertain objectives, *Proceedings of IEEE International Conference on Evolutionary Multi-criterion Optimization*, pp. 314–328.

Teimouri, M., Zaretalab, A., Niaki, S. T. A., and Sharifi, M. (2016). An efficient memory-based electromagnetism-like mechanism for the redundancy allocation problem, *Applied Soft Computing* **38**, pp. 423–436.

Thiele, L., Miettinen, K., Korhonen, P. J., and Molina, J. (2014). A preference-based evolutionary algorithm for multi-objective optimization, *Evolutionary Computation* **17**, 3, pp. 411–436.

Tian, Y., Zhang, X., Cheng, R., and Jin, Y. (2016). A multi-objective evolutionary algorithm based on an enhanced inverted generational distance metric, *Proceedings of the IEEE Congress on Evolutionary Computation*, pp. 5222–5229.

Tillman, F. A., Hwuang, C. L., and Kuo, W. (1980). *Optimization of System Reliability* (Marcel Dekker, New York, United States).

Tomaino, V., Arulselvan, A., Veltri, P., and Pardalos, P. M. (2012). *Studying Connectivity Properties in Human Protein-Protein Interaction Network in Cancer Pathway* (Springer, United States).

Ventresca, M. (2012). Global search algorithms using a combinatorial unranking-based problem representation for the critical node detection problem, *Computers and Operations Research* **39**, 11, pp. 2763–2775.

Ventresca, M. and Aleman, D. (2014). A randomized algorithm with local search for containment of pandemic disease spread, *Computers and Operations Research* **48**, 7, pp. 11–19.

Ventresca, M. and Aleman, D. (2015). Efficiently identifying critical nodes in large complex networks, *Computational Social Networks* **2**, 1, p. 6.

Ventresca, M., Harrison, K. R., and Ombuki-Berman, B. M. (2018). The bi-objective critical node detection problem, *European Journal of Operational Research* **265**, 3.

Veremyev, A., Boginski, V., and Pasiliao, E. L. (2014a). Exact identification of critical nodes in sparse networks via new compact formulations, *Optimization Letters* **8**, 4, pp. 1245–1259.

Veremyev, A., Prokopyev, O. A., and Pasiliao, E. L. (2014b). An integer programming framework for critical elements detection in graphs, *Journal of Combinatorial Optimization* **28**, 1, pp. 233–273.

Walteros, J. L. and Pardalos, P. M. (2012). Selected topics in critical element detection, *Springer Optimization and Its Applications* **71**, pp. 9–26.

Walteros, J. L., Veremyev, A., Pardalos, P. M., and Pasiliao, E. L. (2018). Detecting critical node structures on graphs: A mathematical programming approach, *Networks*.

Wang, G. H., Chen, J., Cai, T., and Xin, B. (2013). Decomposition-based multi-objective differential evolution particle swarm optimization for the design of a tubular permanent magnet linear synchronous motor, *Engineering Optimization* **45**, 9, pp. 1107–1127.

Wang, H., Jiao, L., and Yao, X. (2015). Two_arch2: An improved two-archive algorithm for many-objective optimization, *IEEE Transactions on Evolutionary Computation* **19**, 4, pp. 524–541.

Wang, H. and Yao, X. (2013). Corner sort for Pareto-based many-objective optimization, *IEEE Transactions on Cybernetics* **44**, 1, pp. 92–102.

Wang, R., Xiong, J., Ishibuchi, H., Wu, G., and Zhang, T. (2017). On the effect of reference point in MOEA/D for multi-objective optimization, *Applied Soft Computing* **58**, pp. 25–34.

Wang, Y., Cai, Z., Guo, G., and Zhou, Y. (2007). Multi-objective optimization and hybrid evolutionary algorithm to solve constrained optimization problems, *IEEE Transactions on Systems, Man and Cybernetics, Part B: Cybernetics* **37**, 3, pp. 560–575.

Wang, Z. T., Guo, J. S., Zheng, M. F., and Yang, Y. S. (2014). A new approach for uncertain multiobjective programming problem based on

P_E principle, *Journal of Industrial and Management Optimization* **11**, 1, pp. 13–26.

While, L., Hingston, P., Barone, L., and Huband, S. (2006). A faster algorithm for calculating hypervolume, *IEEE Transactions on Evolutionary Computation* **10**, 1, pp. 29–38.

Wu, L., Wang, H. Y., Lu, F. X., and Jia, P. F. (2008). An anytime algorithm based on modified GA for dynamic weapon-target allocation problem, *Proceedings of IEEE Congress on Computational Intelligence*, pp. 2020–2025.

Xin, B., Chen, J., Peng, Z. H., Dou, L. H., and Zhang, J. (2011). An efficient rule-based constructive heuristic to solve dynamic weapon-target assignment problem, *IEEE Transactions on Systems, Man and Cybernetics, Part A: Systems and Humans* **41**, 3, pp. 598–606.

Xin, B., Chen, J., Zhang, J., Dou, L. H., and Peng, Z. H. (2010). Efficient decision makings for dynamic weapon-target assignment by virtual permutation and Tabu search heuristics, *IEEE Transactions on Systems, Man and Cybernetics, Part C* **40**, 6, pp. 649–662.

Xin, B., Chen, J., Zhang, J., Fang, H., and Peng, Z. H. (2012). Hybridizing differential evolution and particle swarm optimization to design powerful optimizers: A review and taxonomy, *IEEE Transactions on Systems, Man and Cybernetics, Part C: Applications and Reviews* **42**, 5, pp. 744–767.

Xiong, P., Jirutitijaroen, P., and Singh, C. (2017). A distributionally robust optimization model for unit commitment considering uncertain wind power generation, *IEEE Transactions on Power Systems* **32**, 1, pp. 39–49.

Xu, L., Yang, Y. W., and Li, Y. P. (2014). Resource allocation of limited feedback in clustered wireless mesh networks, *Wireless Personal Communications* **75**, 2, pp. 901–913.

Yuan, Y., Xu, H., Wang, B., and Yao, X. (2016). A new dominance relation-based evolutionary algorithm for many-objective optimization, *IEEE Transactions on Evolutionary Computation* **20**, 1, pp. 16–37.

Zapotecas-Martínez, S., Derbel, B., Liefooghe, A., Brockhoff, D., Aguirre, H. E., and Tanaka, K. (2015a). Injecting CMA-ES into MOEA/D, *Proceedings of the 2015 Annual Conference on Genetic and Evolutionary Computation*, pp. 783–790.

Zapotecas-Martínez, S., Derbel, B., Liefooghe, A., Aguirre, H. E., and Tanaka, K. (2015b). Geometric Differential Evolution in MOEA/D: A Preliminary Study, in *Advances in Artificial Intelligence and Soft Computing* (Springer International Publishing, Cham), pp. 364–376.

Zhang, H., Zhou, A., Song, S., Zhang, Q., Gao, X., and Zhang, J. (2016). A self-organizing multiobjective evolutionary algorithm, *IEEE Transactions on Evolutionary Computation* **20**, 5, pp. 792–806.

Zhang, Q. and Li, H. (2007). MOEA/D: A multiobjective evolutionary algorithm based on decomposition, *IEEE Transactions on Evolutionary Computation* **11**, 6, pp. 712–731.

Zhang, Q., Liu, W., and Li, H. (2009a). The performance of a new version of MOEA/D on CEC09 unconstrained MOP test instances, Technical report,

CES-491, School of Computer Science and Electronic Engineering, University of Essex, Colchester, United Kingdom.

Zhang, Q., Liu, W., and Virginas, B. (2010). Expensive multiobjective optimization by MOEA/D with Gaussian process model, *IEEE Transactions on Evolutionary Computation* **14**, 3, pp. 456–474.

Zhang, Q., Zhou, A., Zhao, S., Suganthan, P. N., Liu, W., and Tiwari, S. (2009b). Multiobjective optimization test instances for the CEC 2009 special session and competition, Technical report, CES-487, School of Computer Science and Electronic Engineering, University of Essex, Colchester, United Kingdom.

Zhang, W. and Reimann, M. (2014). A simple augmented ε-constraint method for multi-objective mathematical integer programming problems, *European Journal of Operational Research* **234**, 1, pp. 15–24.

Zhang, L. M., Tang, L. H., and Lei, Y. (2017). Controller area network node reliability assessment based on observable node information, *Frontiers of Information Technology & Electronic Engineering*, Vol. 18, no. 5 pp. 615–626.

Zhou, A., Qu, B., Li, H., Zhao, S., Suganthan, P. N., and Zhang, Q. (2011). Multiobjective evolutionary algorithms: A survey of the state of the art, *Swarm and Evolutionary Computation* **1**, 1, pp. 32–49.

Zhou, A. and Zhang, Q. (2015). Are all the subproblems equally important? Resource allocation in decomposition based multiobjective evolutionary, *IEEE Transactions on Evolutionary Computation* **20**, 1, pp. 52–64.

Zhou, A., Zhang, Q., Jin, Y., Tsang, E., and Okabe, T. (2005). A model-based evolutionary algorithm for bi-objective optimization, *Proceedings of the IEEE Congress on Evolutionary Computation*, pp. 2568–2575.

Zhou, T., Fu, Z., and Wang, B. (2006). Epidemic dynamics on complex networks, *Progress in Natural Science: Materials International* **16**, 5, pp. 452–457.

Zhu, X. and Kuo, W. (2014). Importance measures in reliability and mathematical programming, *Annals of Operations Research* **212**, 1, pp. 241–267.

Zitzler, E., Deb, K., and Thiele, L. (2000). Comparison of multiobjective evolutionary algorithms: Empirical results, *Evolutionary Computation* **8**, 2, pp. 173–195.

Zitzler, E., Laumanns, M., and Thiele, L. (2001). SPEA2: Improving the strength Pareto evolutionary algorithm, Technical report 103, Swiss Federal Institute of Technolo, Switzerland.

Zitzler, E. and Thiele, L. (1999). Multiobjective evolutionary algorithm: A comparative case study and strength Pareto approach, *IEEE Transactions on Evolutionary Computation* **3**, 4, pp. 257–271.

Zitzler, E., Thiele, L., and Laumanns, M. (2003). Performance assessment of multiobjective optimizers: An analysis and review, *IEEE Transactions on Evolutionary Computation* **7**, 2, pp. 117–132.

Zitzler, Z. and Kunzli, S. (2004). Indicator-based selection in multiobjective search, *Parallel Problem Solving from Nature—PPSN VIII: Lecture Notes in Computer Science*, Vol. 3242, pp. 832–842.

Index